CAMBRIDGE TRACTS IN MATHEMATICS

General Editors
B. BOLLOBAS, H. HALBERSTAM & C. T. C. WALL

91 *Fibrewise topology*

I. M. JAMES

Savilian Professor of Geometry
Mathematical Institute, University of Oxford

Fibrewise topology

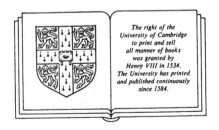

The right of the
University of Cambridge
to print and sell
all manner of books
was granted by
Henry VIII in 1534.
The University has printed
and published continuously
since 1584.

CAMBRIDGE UNIVERSITY PRESS

Cambridge

New York   Port Chester

Melbourne   Sydney

CAMBRIDGE UNIVERSITY PRESS
Cambridge, New York, Melbourne, Madrid, Cape Town, Singapore, São Paulo, Delhi

Cambridge University Press
The Edinburgh Building, Cambridge CB2 8RU, UK

Published in the United States of America by Cambridge University Press, New York

www.cambridge.org
Information on this title: www.cambridge.org/9780521360906

First published 1989
This digitally printed version 2008

A catalogue record for this publication is available from the British Library

Library of Congress Cataloguing in Publication data
James, I. M. (Ioan Mackenzie), 1928–
Fibrewise topology / I.M. James.
p.   cm.   – – (Cambridge tracts in mathematics ; 91)
Bibliography: p.
Includes index.
ISBN 0 521 36090 0
1. Fiberings (Mathematics)   2. Topology.   I. Title.   II. Series.
QA612.6.J36 1988
514'.224 – – dc19   88-10296 CIP

ISBN 978-0-521-36090-6 hardback
ISBN 978-0-521-08925-8 paperback

Contents

Contents

Preface

The aim of this book is to promote the fibrewise viewpoint, particularly in topology. I set out to show that by adopting this viewpoint systematically two desirable objectives can be achieved. First one can reorganize certain portions of the literature and see some well-known material in a new and clearer light. Secondly one is led on to further developments which might not have suggested themselves otherwise.

The fibrewise viewpoint is standard in the theory of fibre bundles, and some of the ideas in this volume originated in work on the foundations of that theory. However, it has been recognized only recently that the same viewpoint is also of great value in other theories, such as general topology. In fact there appear to be fibrewise versions of many of the major concepts of general topology, some of which can already be found in the literature. Having established this I go on to show that there are also fibrewise versions of uniform theory and of homotopy theory. The latter may be well known but has not, as far as I am aware, been written up before. There are important links between fibrewise topology and equivariant topology, a subject which has seen great research activity over the past ten or twenty years. I predict a growing interest in research on fibrewise topology and hope that the present volume will help to provide a foundation for such work.

In fibrewise topology we work over a topological base space B, say. When B is a point-space the theory reduces to that of ordinary topology. A fibrewise topological space over B is just a topological space X together with a continuous projection $p: X \to B$. For example the topological product $B \times T$, for any topological space T, can be regarded as a fibrewise topological space over B, using the first projection, as can any subspace of $B \times T$. For another example suppose that X is a G-space, where G is a topological group; then X can be regarded as a fibrewise topological space over the orbit space X/G.

For any fibrewise topological space over B the inverse images of the points of B, with respect to the projection, are called fibres. We do not require fibres to be homeomorphic, as they are in fibre bundle theory. Although for some results it is necessary for the projection to be surjective (we use the term fibrewise non-empty) in most results the possibility of empty fibres is not excluded.

There are fibrewise versions of many of the important concepts of topology. In some cases all that is necessary is to say that a fibrewise topological space has a particular fibrewise property if each of its fibres has that property in its ordinary form, but in most cases this is not sufficient. One must expect the topology of the base space to play a role.

For example the fibrewise topological space X over B is fibrewise discrete if the projection $p: X \to B$ is a local homeomorphism. For another example X is fibrewise compact if the projection p is proper, in the sense of Bourbaki [9], i.e. if the fibres of X are compact and the projection is closed.

There are five chapters altogether in this volume. Some background in general topology is desirable for the first chapter and rather more for the second. Some previous knowledge of uniform theory is necessary for the third chapter and of homotopy theory for the fourth. The last chapter is miscellaneous in character. However, the non-specialist should not experience any difficulty with the exposition since I have tried to keep it as self-contained as is reasonable for a work of this type. Sets of exercises will be found at the end of each chapter and the volume ends with a note on the literature and a list of references.

Many of the results we shall be proving are fibrewise versions of results which will probably be well known to the reader. If it was just a matter of taking any well-organized account of the ordinary theory and inserting the word 'fibrewise' wherever appropriate there would be little point to the exercise, but there is much more to it than that. First, of course, one has to select suitable fibrewise versions of the various definitions. Often there is some choice: different definitions may work almost equally well. In that situation, of course, one prefers the simplest and most general. In fact some of the definitions used here also appear elsewhere in the literature, for example in the work of Dyckhoff [21], [22], Johnstone [36], [37], Lever [42], [43] and Niefield [49], although of course there tend to be differences of terminology.

Once the appropriate definitions have been chosen it is then moderately straightforward to write down fibrewise versions of the various results in which they are involved. However, one has to be careful. For example in the ordinary version of pointed topology there is seldom any need to assume that the basepoint is a closed set. In the fibrewise version, however, the basepoint is replaced by a section and it is almost always necessary for the section to be closed. To give another example, in the ordinary theory the space of continuous functions from a topological space into a Hausdorff space is itself a Hausdorff space, with the compact-open topology. The corresponding result in the fibrewise theory is false. Again there is no obvious fibrewise version of Urysohn's lemma nor of the Tietze theorem.

Ideally, of course, one would like to have a systematic way of producing fibrewise versions of the definitions and propositions of ordinary topology. Some hope of this is provided by the link between fibrewise topology and topos theory, referred to by Lever [42], [43] and Johnstone [36], [37] amongst others. Unfortunately this approach has several drawbacks, of varying degrees of seriousness. First of all the base space has to satisfy a mild separation condition. Next the method only applies to fibrewise topological spaces where the fibrewise topology is coarser than a fibrewise discrete topology. Thirdly, all the concepts involved must be 'localic' in character, i.e. capable of being formulated in a pointless form. Finally the logical structure of each proposition should be constructivist in nature (although as Lever [43] points out this difficulty can be overcome by using many-valued logic). The method is certainly suggestive, nevertheless, and seems to be of most value where fibrewise compactifications are concerned. Unfortunately it tends to lead to unnecessarily complicated definitions and to propositions encumbered with unnecessary restrictions. Incidentally there is also a theory of uniform locales, due to Isbell [29], Kirwan [41] and Pultr [53], which can be used in a similar fashion although the results obtained seem more related to the Dauns-Hofmann theory of fields of uniform spaces than to the theory of fibrewise uniform spaces described in Chapter III.

An analogy with vector bundle theory may help to clarify these points. At the most basic level it is more or less true that valid results about vector bundles can be obtained from the corresponding results about vector spaces by inserting the word fibrewise in the appropriate places. However, it quickly becomes necessary to take seriously the interaction between the vector space structure of the fibre and the topology of the base. In fact, this is what gives the subject its interest. In our situation, of course, it is the interaction between the *topology* of the fibre and the topology of the base which gives the subject its interest.

On the question of terminology I have, in this volume, used the term 'fibrewise' throughout. The effect is somewhat monotonous, I am afraid, but experience shows that to compromise on this point is liable to cause confusion. Once the fibrewise viewpoint is more widely appreciated it will, I hope, seem reasonable to omit the term in situations where there could not easily be any misunderstanding. At the present stage of development, however, it seems safest always to put it in.

My interest in fibrewise topology goes back many years. I first made my ideas public at a conference in Newfoundland in 1983, which was attended by André Joyal and Miles Tierney; their encouragement and comments helped me greatly. Parts of my book [32] on *General Topology and Homotopy Theory* incorporate those ideas, and my 1984 article in [33] is essentially a summary of what I knew at that time. Since then I have developed much additional material, improving what was there before and correcting errors. For instance, the chapter on fibrewise uniform spaces is based on my Presidential Address [35] to the London Mathematical Society which in turn is derived from an earlier note [34] on the subject.

Among those who have been kind enough to assist me at different stages in the preparation of this work I would particularly like to mention Peter Johnstone, who commented on a first draft, and Alan Pears, who read through the final text. It has also been very helpful to be able to discuss the theory with David Lever who read versions of the text. Two of my research students, Andrew Cook and Zhang Ping, have become interested in fibrewise topology and have made useful suggestions. Finally I would especially like to thank David Lever and Alan Pears for most valuable assistance at the proof stage.

Oxford, 1988   I. M. James

I
Basic fibrewise topology

1. Fibrewise topological spaces

To begin with we work in the category of fibrewise sets over a given set, called the *base set*. If the base set is denoted by B then a *fibrewise set* over B consists of a set X together with a function $p: X \to B$, called the *projection*. For each point b of B the *fibre* over b is the subset $X_b = p^{-1}(b)$ of X; fibres may be empty since we do not require p to be surjective. Also for each subset B' of B we regard $X_{B'} = p^{-1}B'$ as a fibrewise set over B' with the projection determined by p. The alternative notation $X \,|\, B'$ is sometimes convenient.

We regard B as a fibrewise set over itself using the identity as projection. Moreover we regard the Cartesian product $B \times T$, for any set T, as a fibrewise set over B using the first projection.

Let X be a fibrewise set over B with projection p. Then X' is a fibrewise set over B with projection $p\alpha$ for each set X' and function $\alpha: X' \to X$; in particular X' is a fibrewise set over B with projection $p \,|\, X'$ for each subset X' of X. Also X is a fibrewise set over B' with projection βp for each set B' and function $\beta: B \to B'$; in particular X is a fibrewise set over B' with projection given by p for each superset B' of B.

Fibrewise sets over B constitute a category with the following definition of morphism. If X and Y are fibrewise sets over B, with projections p and q, respectively, a function $\phi: X \to Y$ is said to be *fibrewise* (or *fibre-preserving*) if $q\phi = p$, in other words if $\phi X_b \subset Y_b$ for each point b of B. Note that a fibrewise function $\phi: X \to Y$ over B determines, by restriction, a fibrewise function $\phi_{B'}: X_{B'} \to Y_{B'}$ over B' for each subset B' of B. The restriction $\phi_b: X_b \to Y_b$ of ϕ to the fibre over the point b of B can be regarded as an ordinary function.

The equivalences in the category of fibrewise sets over B are called *fibrewise equivalences*. If ϕ, as above, is a fibrewise equivalence over B then $\phi_{B'}$ is a fibrewise equivalence over B' for each subset B' of B, in particular ϕ_b is an equivalence for each point b of B. Conversely if ϕ is a fibrewise function and ϕ_b is an equivalence for each b then ϕ is a fibrewise equivalence. If X is fibrewise equivalent to $B \times T$, for some set T, we say that X is *trivial*, as a fibrewise set over B.

Given an indexed family $\{X_r\}$ of fibrewise sets over B the *fibrewise product* $\prod_B X_r$ is defined, as a fibrewise set over B, and comes equipped with the family of fibrewise projections

$$\pi_r : \prod_B X_r \to X_r.$$

Specifically the fibrewise product is defined as the subset of the ordinary product $\prod X_r$, in which the fibres are the products of the corresponding fibres of the factors X_r. The fibrewise product is characterized by the following Cartesian property: for each fibrewise set X over B the fibrewise functions

$$\phi : X \to \prod_B X_r$$

correspond precisely to the families of fibrewise functions $\{\phi_r\}$, with $\phi_r = \pi_r \phi : X \to X_r$. For example if $X_r = X$ for each index r the *diagonal* $\Delta : X \to \prod_B X$ is defined so that $\pi_r \Delta = \mathrm{id}_X$ for each r.

If $\{X_r\}$ is as before, the *fibrewise coproduct* $\coprod_B X_r$ is also defined, as a fibrewise set over B, and comes equipped with the family of fibrewise insertions

$$\sigma_r : X_r \to \coprod_B X_r.$$

Specifically the fibrewise coproduct coincides, as a set, with the ordinary coproduct (disjoint union), the fibres being the coproducts of the corresponding fibres of the summands X_r. The fibrewise coproduct is characterized by the following co-Cartesian property: for each fibrewise set X over B the fibrewise functions

$$\psi : \coprod_B X_r \to X$$

correspond precisely to the families of fibrewise functions $\{\psi_r\}$, where $\psi_r = \psi \sigma_r : X_r \to X$. For example if $X_r = X$ for each index r the *codiagonal* $\nabla : \coprod_B X \to X$ is defined so that $\nabla \sigma_r = \mathrm{id}_X$ for each r.

Let $\{\phi_r\}$ be a family of fibrewise functions, where $\phi_r : X_r \to X'_r$ and X_r and X'_r are fibrewise sets over B. Then the fibrewise product

$$\prod \phi_r : \prod_B X_r \to \prod_B X'_r$$

is defined, so that $\pi'_r(\prod \phi_r) = \phi_r \pi_r$ for each r, and the fibrewise coproduct

$$\coprod \phi_r : \coprod_B X_r \to \coprod_B X'_r$$

is defined, so that $(\coprod \phi_r)\sigma_r = \sigma'_r\phi_r$ for each r. Note that if ϕ_r is a fibrewise equivalence for each r then $\prod \phi_r$ and $\coprod \phi_r$ are also fibrewise equivalences.

The notations $X \times_B Y$ and $X +_B Y$ are used for the fibrewise product and coproduct in the case of a family $\{X, Y\}$ of two fibrewise sets and similarly for finite families generally.

Now suppose that B is a topological space. By a *fibrewise topology* on a fibrewise set X over B I mean any topology on X for which the projection p is continuous. The coarsest such topology is the topology induced by p, in which the open sets of X are precisely the inverse images of the open sets of B; this is called the *fibrewise indiscrete topology*. A *fibrewise topological space* over B is defined to be a fibrewise set over B with a fibrewise topology.

We regard B as a fibrewise topological space over itself using the identity as projection. Also, we regard the topological product $B \times T$, for any topological space T, as a fibrewise topological space over B, using the first projection, and similarly for any subspace of $B \times T$. By a *section* of a fibrewise topological space we mean a continuous right inverse of the projection.

Definition (1.1). *The fibrewise function $\phi: X \to Y$ is fibrewise constant where X and Y are fibrewise topological over B, if $\phi = tp$ for some section $t: B \to Y$.*

A continuous fibrewise function may be constant on each fibre and yet not fibrewise constant. For example let B be a non-discrete space and let $X = Y$ be the same set with the discrete topology and the identity as projection; then the identity function is not fibrewise constant.

Let X be fibrewise topological over B with projection p. Then X' is fibrewise topological over B with projection $p\alpha$ for each topological space X' and continuous function $\alpha: X' \to X$; in particular X' is fibrewise topological over B with projection $p|X'$ for each subspace X' of X. Also X is fibrewise topological over B' with projection βp for each topological space B' and continuous function $\beta: B \to B'$; in particular X is fibrewise topological over B' with projection given by p for each superspace B' of B.

Fibrewise topological spaces over a base form a category in which the morphisms are continuous fibrewise functions. If $\phi: X \to Y$ is a fibrewise function, where X and Y are fibrewise topological over B, then ϕ is necessarily continuous whenever Y has the fibrewise indiscrete topology.

Note that a continuous fibrewise function $\phi: X \to Y$ over B determines, by restriction, a continuous fibrewise function $\phi_{B'}: X_{B'} \to Y_{B'}$ over B' for each subspace B' of B. In the case in which B' reduces to a point the restriction can be regarded as an ordinary continuous function.

The equivalences in the category are called *fibrewise topological equivalences*. If ϕ, as above, is a fibrewise topological equivalence then so is $\phi_{B'}$ for each subspace B' of B. In particular ϕ_b is a homeomorphism for each point b of B, but this necessary condition for a fibrewise topological equivalence is not in general sufficient. For example take $Y = B$, and take $X = B$ with a finer topology; the identity function is continuous and homeomorphic on fibres but is not a homeomorphism.

If X is fibrewise topologically equivalent to $B \times T$, for some topological space T, we say that X is *trivial*, as a fibrewise topological space over B. *Local triviality* in the same sense is defined similarly.

Let $\phi: X \to Y$ be a fibrewise function where X is a fibrewise set and Y is a fibrewise topological space over B. We can give X the induced topology, in the ordinary sense, and this is necessarily a fibrewise topology. We may refer to it, therefore, as the *induced fibrewise topology* and note the following characterization.

Proposition (1.2). *Let $\phi: X \to Y$ be a fibrewise function, where Y is fibrewise topological over B and X has the induced fibrewise topology. Then for each fibrewise topological Z, a fibrewise function $\psi: Z \to X$ is continuous if and only if the composition $\phi\psi: Z \to Y$ is continuous.*

Similarly in the case of families $\{\phi_r\}$ of fibrewise functions, where $\phi_r: X \to Y$, with Y_r fibrewise topological over B for each r. In particular, given a family $\{X_r\}$ of fibrewise topological spaces over B, the fibrewise topological product $\prod_B X_r$ is defined to be the fibrewise product with the fibrewise topology induced by the family of projections. Then for each fibrewise topological space Z over B a fibrewise function $\theta: Z \to \prod_B X_r$ is continuous if and only if each of the fibrewise functions $\pi_r \theta: Z \to X_r$ is continuous. For example when $X_r = X$ for each index r we see that the diagonal $\Delta: X \to \prod_B X$ is continuous.

Again if $\{X_r\}$ is a family of fibrewise topological spaces over B the fibrewise topological coproduct $\coprod_B X_r$ is just the fibrewise coproduct at the set-theoretic level with the ordinary coproduct topology. Thus if $\{\psi_r\}$ is a family of continuous fibrewise functions, where $\psi_r: X_r \to X$ and X is fibrewise topological over B then the fibrewise function $\psi: \coprod_B X_r \to X$ given by $\psi\sigma_r = \psi_r$ for each index r is continuous. For example when $X_r = X$ for each index r we see that the codiagonal $\nabla: \coprod_B X \to X$ is continuous.

Finally let $\{\phi_r\}$ be a family of continuous fibrewise functions, where $\phi_r: X_r \to X'_r$ and X_r and X'_r are fibrewise topological over B for each index r. Then the fibrewise product and coproduct

$$\prod_B \phi_r: \prod_B X_r \to \prod_B X'_r, \qquad \coprod_B \phi_r: \coprod_B X_r \to \coprod_B X'_r$$

are continuous. Note that if ϕ_r is a fibrewise topological equivalence for each index r then $\prod_B \phi_r$ and $\coprod_B \phi_r$ are also fibrewise topological equivalences.

In fibrewise topology the term neighbourhood is used in precisely the same sense as it is in ordinary topology, but the terms *fibrewise basic* and *subbasic* neighbourhood may need some explanation. Thus let X be fibrewise topological over B. If x is a point of X_b ($b \in B$) I describe a family Γ of neighbourhoods of x in X as *fibrewise basic* if for each neighbourhood U of x we have $X_W \cap V \subset U$, for some member V of Γ and neighbourhood W of b in B. For example, in the case of the topological product $B \times T$, where T is a topological space, the family of Cartesian products $B \times N$, where N runs through the neighbourhoods of t, is fibrewise basic for (b, t). The term fibrewise subbasic neighbourhood is used similarly.

It may be convenient to mention here just a few results which apply to a fibrewise function $\phi: X \to Y$, where X and Y are fibrewise topological over B. Recall that ϕ is continuous if for each point $x \in X_b$, $b \in B$, the inverse image of each neighbourhood of $\phi(x)$ is a neighbourhood of x; it is sufficient if this condition is satisfied for fibrewise subbasic neighbourhoods of $\phi(x)$, where appropriate. Also recall that ϕ is open if for each point $x \in X_b$, $b \in B$, the direct image of each neighbourhood of x is a neighbourhood of $\phi(x)$; in this case it is sufficient if the condition is satisfied for fibrewise basic neighbourhoods. When ϕ is injective it is even sufficient if the condition is satisfied for fibrewise subbasic neighbourhoods. We also make use of the less familiar†

Proposition (1.3). *Let $\phi: X \to Y$ be an open and closed fibrewise surjection, where X and Y are fibrewise topological over B. Let $\alpha: X \to \mathbb{R}$ be a continuous real-valued function which is fibrewise bounded above, in the sense that α is bounded above on each fibre of X. Then $\beta: Y \to \mathbb{R}$ is continuous, where*

$$\beta(\eta) = \sup_{\xi \in \phi^{-1}(\eta)} \alpha(\xi).$$

Let us show that β is continuous at the point y of Y. Let $\varepsilon > 0$. Since α is continuous at each point $x \in \phi^{-1}(y)$ there exists a neighbourhood U_x of x such that

$$\alpha(x) - \varepsilon/2 < \alpha(\xi) < \alpha(x) + \varepsilon/2$$

for all $\xi \in U_x$. The union U of these neighbourhoods U_x, for $x \in \phi^{-1}(y)$, is a neighbourhood of $\phi^{-1}(y)$. Since ϕ is closed there exists a neighbourhood V

---

† This useful result occurs as an exercise in Dugundi [20]: I do not know where it originated.

of y such that $\phi^{-1}V \subset U$. Now if $\eta \in V$ we have $\beta(\eta) < \alpha(\xi) + \varepsilon/2$, for some $\xi \in \phi^{-1}(\eta)$, and then $\alpha(\xi) < \alpha(x) + \varepsilon/2$, for some $x \in \phi^{-1}(y)$, so that $\beta(\eta) < \alpha(x) + \varepsilon \leqslant \beta(y) + \varepsilon$. On the other hand $\beta(y) - \varepsilon/2 < \alpha(x)$, for some $x \in \phi^{-1}(y)$, and then $\alpha(x) - \varepsilon/2 < \alpha(\xi)$, for all $\xi \in U_x$, so that $\beta(y) - \varepsilon < \alpha(\xi) \leqslant \beta\phi(\xi)$. Thus if $\eta \in V \cap \phi U_x$ then

$$\beta(y) - \varepsilon < \beta(\eta) < \beta(y) + \varepsilon.$$

But ϕU_x is open, since ϕ is open, and so $V \cap \phi U_x$ is a neighbourhood of y. Therefore β is continuous at y, which proves (1.3).

By a *proper fibrewise function* we mean a fibrewise function which is proper in the ordinary sense, as in Bourbaki [9]. A convenient characterization is given in

Proposition (1.4). *Let* $\phi : X \to Y$ *be a fibrewise function, where X and Y are fibrewise topological over B. Then ϕ is proper if and only if*

$$\phi \times \mathrm{id} : X \times_B T \to Y \times_B T$$

is closed for all fibrewise topological T.

In what is to follow we shall be studying various classes of fibrewise topological spaces, ranging from the universal class, consisting of all such spaces, to the class which consists of the base space alone. To be significant, for our purposes, a class must in any case be invariant, in the sense of fibrewise topological equivalence. Suppose that we have an invariant class of fibrewise topological spaces, defined for every base space. When the base space is a point this will reduce to an invariant class of topological spaces, for example the class of discrete spaces or compact spaces. When the base space is a given space B, say, we naturally use the terminology suggested by the special case, for example the class of fibrewise discrete spaces or fibrewise compact spaces over B. However this is just a matter of convenience and is not meant to imply that there is a unique way to generalize an invariant property P of topological spaces to an invariant property fibrewise P of fibrewise topological spaces.

Fibrewise topological spaces where the projection is closed or open have some interesting properties. Although most of the following results are just special cases of well-known results in ordinary topology, or are otherwise obvious, there are a few where this is not so and for these full proofs are given.

Definition (1.5). *The fibrewise topological space X over B is* fibrewise closed *if the projection p is closed.*

For example, trivial fibrewise spaces with compact fibre are fibrewise closed. Also G-spaces, where G is compact, are fibrewise closed over their orbit spaces.

Proposition (1.6). *Let $\phi: X \to Y$ be a closed fibrewise function, where X and Y are fibrewise topological over B. Then X is fibrewise closed if Y is fibrewise closed.*

Proposition (1.7). *Let X be fibrewise topological over B. Suppose that X_j is fibrewise closed for each member X_j of a finite covering of X. Then X is fibrewise closed.*

Proposition (1.8). *Let X be fibrewise topological over B. Then X is fibrewise closed if and only if for each fibre X_b of X and each neighbourhood U of X_b in X there exists a neighbourhood W of b such that $X_W \subset U$.*

Definition (1.9). *The fibrewise topological space X over B is* fibrewise open *if the projection p is open.*

There is an extensive literature concerning fibrewise open spaces: for some relevant material see Lewis [44]. For example, trivial fibrewise spaces are always fibrewise open. Also G-spaces are always fibrewise open over their orbit spaces.

Proposition (1.10). *Let $\phi: X \to Y$ be an open fibrewise function, where X and Y are fibrewise topological over B. Then X is fibrewise open if Y is fibrewise open.*

Proposition (1.11). *Let $\{X_r\}$ be a finite family of fibrewise open spaces over B. Then the fibrewise topological product $X = \prod_B X_r$ is also fibrewise open.*

In other words the class of fibrewise open spaces is finitely multiplicative. In fact (1.11) remains true for infinite families provided each member of the family is fibrewise non-empty in the sense that the projection is surjective.

Note that if X is fibrewise open then the second projection $\pi_2: X \times_B Y \to Y$ is open for all fibrewise topological Y. We use this in the proof of

Proposition (1.12). *Let $\phi: X \to Y$ be a fibrewise function, where X and Y are fibrewise topological over B. Suppose that the product*

$$\mathrm{id} \times \phi: X \times_B X \to X \times_B Y$$

is open and that X is fibrewise open. Then ϕ itself is open.

For consider the commutative diagram shown below.

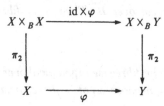

The projection on the left is surjective, while the projection on the right is open, since X is fibrewise open. Therefore $\pi_2(\mathrm{id} \times \phi) = \phi\pi_2$ is open, and so ϕ is open, by (1.10), as asserted.

Our next three results apply equally to fibrewise closed and fibrewise open spaces.

Proposition (1.13). *Let* $\phi: X \to Y$ *be a continuous fibrewise surjection, where X and Y are fibrewise topological over B. If X is fibrewise closed (resp. open) then Y is fibrewise closed (resp. open).*

Proposition (1.14). *Let X be fibrewise topological over B. Suppose that X is fibrewise closed (resp. open) over B. Then $X_{B'}$ is fibrewise closed (resp. open) over B' for each subspace B' of B.*

Proposition (1.15). *Let X be fibrewise topological over B. Suppose that X_{B_j} is fibrewise closed (resp. open) over B_j for each member B_j of an open covering of B. Then X is fibrewise closed (resp. open) over B.*

In fact the last result is also true for locally finite closed coverings.

There are several subclasses of the class of fibrewise open spaces which include many important examples and have interesting properties.

Definition (1.16). *The fibrewise topological space X over B is* locally sliceable *if for each point $x \in X_b$, where $b \in B$, there exists a neighbourhood W of b and a section $s: W \to X_W$ such that $s(b) = x$.*

The condition implies that p is open since if U is a neighbourhood of x in X then $s^{-1}(X_W \cap U) \subset pU$ is a neighbourhood of b in W and hence in B. For example (see [16]) completely regular G-spaces, with G compact Lie, are locally sliceable over their orbit spaces. The class of locally sliceable spaces is finitely multiplicative as stated in

Proposition (1.17). *Let $\{X_r\}$ $(r = 1, \ldots, n)$ be a finite family of locally sliceable spaces over B. Then the fibrewise topological product $X = \prod_B X_r$ is locally sliceable.*

For let $x = (x_r)$ be a point of X_b, where $b \in B$, so that $x_r = \pi_r(x)$ for each index r. Since X_r is locally sliceable there exists a neighbourhood W_r of b and a section $s_r \colon W_r \to X_r \,|\, W_r$ such that $s_r(b) = x_r$. Then the intersection $W = W_1 \cap \cdots \cap W_n$ is a neighbourhood of b and a section $s \colon W \to X_W$ is given by $\pi_r s(\beta) = s_r(\beta)$ for each index r and each point $\beta \in W$.

Proposition (1.18). *Let* $\phi \colon X \to Y$ *be a continuous fibrewise surjection, where X and Y are fibrewise topological over B. If X is locally sliceable then so is Y.*

For let $y \in Y_b$, where $b \in B$. Then $y = \phi(x)$, for some $x \in X_b$. If X is locally sliceable there exists a neighbourhood W of b and a section $s \colon W \to X_W$ such that $s(b) = x$. Then $\phi s \colon W \to Y_W$ is a section such that $s(b) = y$, as required.

Definition (1.19). *The fibrewise topological space X over B is* fibrewise discrete *if the projection p is a local homeomorphism.*

This means, we recall, that for each point b of B and each point x of X_b there exists a neighbourhood V of x in X and a neighbourhood W of b in B such that p maps V homeomorphically onto W; in that case we say that W is evenly covered by V. For example (see §7) properly discontinuous G-spaces, with G discrete, are fibrewise discrete over their orbit spaces. Clearly fibrewise discrete spaces are locally sliceable and hence fibrewise open.

Proposition (1.20). *Let $\{X_r\}$ be a finite family of fibrewise discrete spaces over B. Then the fibrewise topological product $X = \prod_B X_r$ is fibrewise discrete.*

In other words the class of fibrewise discrete spaces is finitely multiplicative. For, given a point $x \in X_b$, there exists for each index r a neighbourhood U_r of $\pi_r(x)$ in X_r, such that the projection $p_r = \pi_r p$ maps U_r homeomorphically onto the neighbourhood $p_r U_r = W_r$ of b. Then the neighbourhood $\prod_B U_r$ of x is mapped homeomorphically by p onto the intersection $W = \bigcap W_r$, which is a neighbourhood of b.

An attractive characterization of fibrewise discrete spaces is given by the following†

---

† I have not been able to find a reference for this result which seems to be less well known than one would have expected. However, a localic version appears in the *Memoir* of Joyal and Tierney [39].

Proposition (1.21). *Let X be a fibrewise topological space over B. Then X is fibrewise discrete if and only if* (i) *X is fibrewise open and* (ii) *the diagonal embedding*

$$\Delta : X \to X \times_B X$$

is open.

For suppose that (i) and (ii) are satisfied. Let $x \in X_b$, where $b \in B$. Then $\Delta(x) = (x, x)$ admits a neighbourhood in $X \times_B X$ which is entirely contained in ΔX. Without real loss of generality we may assume the neighbourhood is of the form $U \times_B U$, where U is a neighbourhood of x in X. Then $p \mid U$ is a homeomorphism. Therefore X is fibrewise discrete.

Conversely suppose that X is fibrewise discrete. We have already seen that X is fibrewise open. To show that Δ is open it is sufficient to show that ΔX is open in $X \times_B X$. So let $x \in X_b$, where $b \in B$, and let U be a neighbourhood of x in X such that $W = pU$ is a neighbourhood of b in B and p maps U homeomorphically onto W. Then $U \times_B U$ is contained in ΔX since if not then there exist distinct $\xi, \xi' \in X_\beta$, where $\beta \in W$ and $\xi, \xi' \in U$, which is absurd.

Open subspaces of fibrewise discrete spaces are also fibrewise discrete. In fact we have

Proposition (1.22). *Let $\phi : X \to Y$ be a continuous fibrewise injection, where X and Y are fibrewise open over B. If Y is fibrewise discrete then so is X.*

For consider the diagram shown below.

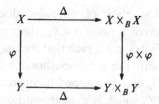

Since ϕ is continuous so is $\phi \times \phi$. Now ΔY is open in $Y \times_B Y$, by (1.21), since Y is fibrewise discrete. So $\Delta X = \Delta \phi^{-1} Y = (\phi \times \phi)^{-1} \Delta Y$ is open in $X \times_B X$. Thus (1.22) follows from (1.21).

Proposition (1.23). *Let $\phi: X \to Y$ be an open fibrewise surjection, where X and Y are fibrewise open over B. If X is fibrewise discrete then so is Y.*

For in the above diagram, with these fresh hypotheses on ϕ, if X is fibrewise discrete then ΔX is open in $X \times_B X$, by (1.21), and so $\Delta Y = \Delta \phi X = (\phi \times \phi) \Delta X$ is open in $Y \times_B Y$. Thus (1.23) follows from (1.21) again.

Proposition (1.24). *Let $\phi, \psi: X \to Y$ be continuous fibrewise functions, where X is fibrewise topological and Y is fibrewise discrete over B. Then the coincidence set $K(\phi, \psi)$ of ϕ and ψ is open in X.*

For the coincidence set is precisely $\Delta^{-1}(\phi \times \psi)^{-1} \Delta Y$, where

$$X \xrightarrow{\Delta} X \times_B X \xrightarrow{\phi \times \psi} Y \times_B Y \xleftarrow{\Delta} Y.$$

Hence (1.24) follows at once from (1.21). In particular take $X = Y$, take $\phi = \mathrm{id}_X$ and take $\psi = sp$, where s is a section; we conclude that s is an open embedding when X is fibrewise discrete.

Proposition (1.25). *Let $\phi: X \to Y$ be a continuous fibrewise function, where X is fibrewise open and Y is fibrewise discrete over B. Then the fibrewise graph*

$$\Gamma: X \to X \times_B Y$$

of ϕ is an open embedding.

Here the fibrewise graph is defined in the same way as the ordinary graph, but with values in the fibrewise product, so that the diagram shown below is commutative.

Since ΔY is open in $Y \times_B Y$, by (1.21), so $\Gamma X = (\phi \times \mathrm{id})^{-1} \Delta Y$ is open in $X \times_B Y$, as asserted.

Note that if X is fibrewise discrete over B then for each point $x \in X_b$, $b \in B$, there exists a neighbourhood W of b such that a unique section $s: W \to X_W$ exists satisfying $s(b) = x$. We may refer to s as *the* section through x.

Definition (1.26). *The fibrewise topological space* X *over* B *is* locally sectionable *if each point* b *of* B *admits a neighbourhood* W *and section* $s: W \to X_W$.

Thus fibrewise non-empty locally sliceable spaces are locally sectionable, for example. However, the converse is false (see Figure 1). In fact, locally sectionable spaces are not necessarily fibrewise open; for example take $X = (-1, 1] \subset \mathbb{R}$ with the natural projection onto $B = \mathbb{R}/\mathbb{Z}$.

Proposition (1.27). *Let* $\{X_r\}$ *be a finite family of locally sectionable spaces over* B. *Then the fibrewise topological product* $\prod_B X_r$ *is locally sectionable.*

In other words the class of locally sectionable spaces is finitely multiplicative.

For, given a point b of B, there exists a neighbourhood W_r of b and a section $s_r: W_r \to X_r | W_r$ for each index r. Since there are only a finite number of indices the intersection W of the neighbourhoods W_r is also a neighbourhood of b, and a section $s: W \to (\prod_B X_r)_W$ is given by $\pi_r s(\beta) = s_r(\beta)$, for $\beta \in W$.

Our last two results apply equally well to each of the above three properties.

Proposition (1.28). *Let* X *be fibrewise topological over* B. *Suppose that* X *is locally sliceable, fibrewise discrete, or locally sectionable over* B. *Then so is* $X_{B'}$ *over* B' *for each open set* B' *of* B.

Figure 1

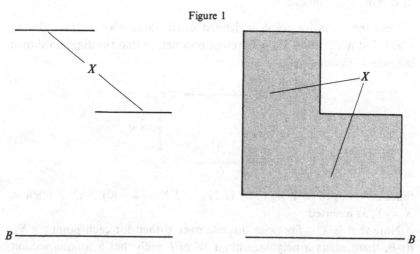

not locally sectionable   locally sectionable but not
locally sliceable

Proposition (1.29). *Let X be fibrewise topological over B. Suppose that X_{B_j} is locally sliceable, fibrewise discrete or locally sectionable over B_j for each member B_j of an open covering of B. Then so is X over B.*

There is an interesting class of fibrewise discrete spaces called overlayings which were introduced by Fox [26] in the context of shape theory (see also Moore [48]). An overlay structure on the fibrewise set X over B is given by a basis $\{W_j\}$ for the topology of B and for each index j a family Γ_j of subsets of X_{W_j}, forming a covering, such that the following two conditions are satisfied. First for each index j each member of Γ_j projects bijectively onto W_j under p. Secondly for each pair of indices j, k the intersection of a member of Γ_j with a member of Γ_k is either empty or projects onto $W_j \cap W_k$. In particular the members of Γ_j, for each index j, form a partition of X_{W_j}. Now the collection of members of the families Γ_j, for all indices j, forms a basis for a topology on X, which is necessarily fibrewise discrete. We call such a fibrewise discrete topology an *overlay structure* and refer to X, with this structure, as an *overlaying* of B. Fox gives examples to show that while every overlaying is a covering, and every finite covering is an overlaying, there exist infinite coverings which do not admit overlay structure.

There are various unsolved problems about fibrewise discrete spaces. For example, let X be fibrewise open over B. Obviously X is fibrewise discrete if there exists a continuous fibrewise injection $X \to B \times T$ for some discrete T. When does the converse hold? Dyckhoff [21] has discussed this problem for finite T.

I do not know whether, in general, a fibrewise set admits a fibrewise discrete topology. However it is not difficult to give examples of different fibrewise discrete topologies on the same fibrewise set which are inequivalent, as fibrewise topologies. For this reason we must be careful not to say *the* fibrewise discrete topology.

2. Fibrewise separation conditions

The purpose of this section is to consider fibrewise versions† of some of the more important separation conditions of ordinary topology, beginning with

Definition (2.1). *Let X be fibrewise topological over B. Then X is* fibrewise

---

† Perhaps it should be pointed out that in the case of a fibrewise topological space over B, with B paracompact, each of these fibrewise separation conditions is equivalent to the corresponding ordinary separation condition on X, as a topological space.

T_0 if whenever $x, x' \in X_b$, where $b \in B$ and $x \neq x'$, either there exists a neighbourhood of x which does not contain x', or vice versa.

In other words, X is fibrewise T_0 if and only if each fibre X_b is T_0. It is easy to check that subspaces of fibrewise T_0-spaces are fibrewise T_0, also the fibrewise topological products of fibrewise T_0 spaces are fibrewise T_0.

Of course one can formulate a fibrewise version of the T_1 condition in a similar fashion, but it turns out that there is no real use for this in what we are going to do. Instead we make some use of another condition which has been considered by Davis [14] amongst others. The condition is that every open set contains the closure of each of its points. This is true for T_1-spaces, of course, and for regular spaces. Thinking of it as a weak form of regularity Davis uses the term R_0-space. Indiscrete spaces are R_0-spaces, for example, although they are not T_0. The fibrewise version of the R_0 condition is as follows.

Definition (2.2). *The fibrewise topological space X over B is fibrewise R_0 if for each point $x \in X_b$, where $b \in B$, and each neighbourhood V of x in X, there exists a neighbourhood W of b in B such that $X_W \cap \overline{\{x\}} \subset V$.*

When the neighbourhoods of x are given by a fibrewise basis it is sufficient if the condition in (2.2) is satisfied for all fibrewise basic neighbourhoods. For example $B \times T$ is fibrewise R_0 for all R_0-spaces T. Note that if X is fibrewise R_0 over B then $X_{B'}$ is fibrewise R_0 over B' for each subspace B' of B.

Subspaces of fibrewise R_0-spaces are fibrewise R_0. In fact we have

Proposition (2.3). *Let $\phi: X \to X'$ be a fibrewise embedding, where X and X' are fibrewise topological over B. If X' is fibrewise R_0 then so is X.*

For let $x \in X_b$, where $b \in B$, and let V be a neighbourhood of x in X. Then $V = \phi^{-1}V'$, where V' is a neighbourhood of $x' = \phi(x)$ in X'. If X' is fibrewise R_0 there exists a neighbourhood W of b such that $X'_W \cap \overline{\{x'\}} \subset V'$. Then

$$X_W \cap \overline{\{x\}} \subset \phi^{-1}(X'_W \cap \overline{\{x'\}}) \subset \phi^{-1}V' = V,$$

and so X' is fibrewise R_0 as asserted.

Proposition (2.4). *Let $\{X_r\}$ be a finite family of fibrewise R_0-spaces over B. Then the fibrewise topological product $X = \prod_B X_r$ is fibrewise R_0.*

In other words the class of fibrewise R_0-spaces is finitely multiplicative
For let $x \in X_b$, where $b \in B$. Consider a neighbourhood $V = \prod_B V_r$ of x in X,
where V_r is a neighbourhood of $\pi_r(x) = x_r$ in X_r for each index r. Since X_r is
fibrewise R_0 there exists a neighbourhood W_r of b in B such that
$(X_r | W_r) \cap \overline{\{x_r\}} \subset V_r$. Then the intersection W of the W_r is a neighbourhood
of b such that $X_W \cap \overline{\{x\}} \subset V$, as required. The same conclusion holds for
infinite fibrewise products provided each of the factors is fibrewise non-
empty.

Proposition (2.5). *Let $\phi : X \to Y$ be a closed continuous fibrewise surjection,
where X and Y are fibrewise topological over B. If X is fibrewise R_0 then
so is Y.*

For let $y \in Y_b$, where $b \in B$, and let V be a neighbourhood of y in Y.
Pick $x \in \phi^{-1}(y)$. Then $U = \phi^{-1}V$ is a neighbourhood of x. If X is fibrewise
R_0 there exists a neighbourhood W of b such that $X_W \cap \overline{\{x\}} \subset U$. Then
$Y_W \cap \phi\overline{\{x\}} \subset \phi U = V$. But $\phi\overline{\{x\}} = \overline{\{\phi(x)\}}$, since ϕ is closed, and so Y is
fibrewise R_0 as asserted.

Definition (2.6). *The fibrewise topological space X over B is fibrewise
Hausdorff if whenever $x, x' \in X_b$, where $b \in B$ and $x \neq x'$, there exist disjoint
neighbourhoods V, V' of x, x' in X.*

For example $B \times T$ is fibrewise Hausdorff for all Hausdorff T. Note that
if X is fibrewise Hausdorff over B then $X_{B'}$ is fibrewise Hausdorff over B' for
each subspace B' of B. In particular the fibres of X are Hausdorff spaces.
However a fibrewise topological space with Hausdorff fibres is not
necessarily Hausdorff: take $X = B$ with B indiscrete.

Proposition (2.7). *The fibrewise topological space X over B is fibrewise
Hausdorff if and only if the diagonal embedding*
$$\Delta : X \to X \times_B X$$
is closed.

As before let $x, x' \in X_b$, where $b \in B$ and $x \neq x'$. If ΔX is closed in $X \times_B X$
then (x, x'), being a point of the complement, admits a fibrewise product
neighbourhood $V \times_B V'$ which does not meet ΔX, and then V, V' are
disjoint neighbourhoods of x, x'. The argument in the reverse direction is
similar.

For example, let Γ be a topological group and let G be a Hausdorff
subgroup of Γ. Then the diagonal is the inverse image of the neutral

element with respect to the division function $d : \Gamma \times_{\Gamma/G} \Gamma \to G$. Therefore the diagonal is closed and so Γ is fibrewise Hausdorff over Γ/G.

Subspaces of fibrewise Hausdorff spaces are fibrewise Hausdorff. In fact we have

Proposition (2.8). *Let* $\phi : X \to Y$ *be a continuous fibrewise injection, where X and Y are fibrewise topological over B. If Y is fibrewise Hausdorff then so is X.*

For let $x, x' \in X_b$, where $b \in B$ and $x \neq x'$. Then $\phi(x), \phi(x') \in Y_b$ are distinct and so, if Y is fibrewise Hausdorff, there exist neighbourhoods V, V' of $\phi(x), \phi(x')$ in Y which are disjoint. Their inverse images $\phi^{-1}V$, $\phi^{-1}V'$ in X are neighbourhoods of x, x' in X which are disjoint. Alternatively (2.7) can be used.

Proposition (2.9). *Let* $\phi, \psi : X \to Y$ *be continuous fibrewise functions, where X and Y are fibrewise topological over B. If Y is fibrewise Hausdorff then the coincidence set* $K(\phi, \psi)$ *is closed in X.*

Recall that the coincidence set is precisely $\Delta^{-1}(\phi \times \psi)^{-1}\Delta Y$, where

$$X \xrightarrow{\ \Delta\ } X \times_B X \xrightarrow{\ \phi \times \psi\ } Y \times_B Y \xleftarrow{\ \Delta\ } Y.$$

Hence (2.9) follows at once from (2.7). In particular take $X = Y$, take $\phi = \mathrm{id}_X$ and take $\psi = sp$, where s is a section; we conclude that s is closed when X is fibrewise Hausdorff.

Proposition (2.10). *Let* $\phi : X \to Y$ *be a continuous fibrewise function, where X and Y are fibrewise topological over B. If Y is fibrewise Hausdorff then the fibrewise graph*

$$\Gamma : X \to X \times_B Y$$

of ϕ is a closed embedding.

This follows similarly from the commutative diagram shown below.

$$
\begin{array}{ccc}
X & \xrightarrow{\ \Gamma\ } & X \times_B Y \\
{\scriptstyle \varphi}\downarrow & & \downarrow{\scriptstyle \varphi \times \mathrm{id}} \\
Y & \xrightarrow{\ \Delta\ } & Y \times_B Y
\end{array}
$$

The class of fibrewise Hausdorff spaces is multiplicative in the following sense.

Proposition (2.11). *Let* $\{X_r\}$ *be a family of fibrewise Hausdorff spaces over* B. *Then the fibrewise topological product* $X = \prod_B X_r$ *is fibrewise Hausdorff.*

For let $x, x' \in X_b$, where $b \in B$ and $x \neq x'$. Then $\pi_r(x) \neq \pi_r(x')$ for some index r. Since X_r is fibrewise Hausdorff there exist neighbourhoods V, V' of $\pi_r(x), \pi_r(x')$ in X_r which are disjoint. Then $\pi_r^{-1}(V), \pi_r^{-1}(V')$ are disjoint neighbourhoods of x, x' in X, as required.

The functional version of the fibrewise Hausdorff condition is stronger than the non-functional version but its properties are fairly similar. Here and elsewhere we use I to denote the closed unit interval $[0, 1]$ in the real line $\mathbb{R}$.

Definition (2.12). *The fibrewise topological space* X *over* B *is fibrewise functionally Hausdorff if whenever* $x, x' \in X_b$, *where* $b \in B$ *and* $x \neq x'$, *there exists a neighbourhood* W *of* b *and a continuous function* $\alpha \colon X_W \to I$ *such that* $\alpha(x) = 0$ *and* $\alpha(x') = 1$.

For example $B \times T$ is fibrewise functionally Hausdorff for all functionally Hausdorff T. Note that if X is fibrewise functionally Hausdorff over B then $X' = X_{B'}$ is fibrewise functionally Hausdorff over B' for each subspace B' of B. In particular the fibres of X are functionally Hausdorff spaces.

Subspaces of fibrewise functionally Hausdorff spaces are fibrewise functionally Hausdorff. In fact we have

Proposition (2.13). *Let* $\phi \colon X \to Y$ *be a continuous fibrewise injection, where* X *and* Y *are fibrewise topological over* B. *If* Y *is fibrewise functionally Hausdorff then so is* X.

Moreover the class of fibrewise functionally Hausdorff spaces is multiplicative, as stated in

Proposition (2.14). *Let* $\{X_r\}$ *be a family of fibrewise functionally Hausdorff spaces over* B. *Then the fibrewise topological product* $X = \prod_B X_r$ *is fibrewise functionally Hausdorff.*

The proofs are similar to those for the corresponding results in the non-functional case and will therefore be omitted.

We now proceed to consider the fibrewise versions of the higher separation conditions, starting with regularity and complete regularity.

Definition (2.15). *The fibrewise topological space X over B is fibrewise regular† if for each point $x \in X_b$, where $b \in B$, and for each neighbourhood V of x in X, there exists a neighbourhood W of b in B and a neighbourhood U of x in X_W such that the closure $X_W \cap \bar{U}$ of U in X_W is contained in V (see Figure 2).*

When the neighbourhoods of x are given by a fibrewise basis it is sufficient if the condition in (2.15) is satisfied for all fibrewise basic neighbourhoods. For example trivial fibrewise spaces with regular fibre are fibrewise regular. Also (see §7) proper G-spaces are fibrewise regular over their orbit spaces. Note that if X is fibrewise regular over B then $X_{B'}$ is fibrewise regular over B' for each subspace B' of B.

Subspaces of fibrewise regular spaces are fibrewise regular. In fact we have

Proposition (2.16). *Let $\phi : X \to X'$ be a fibrewise embedding, where X and X' are fibrewise topological over B. If X' is fibrewise regular then so is X.*

Figure 2

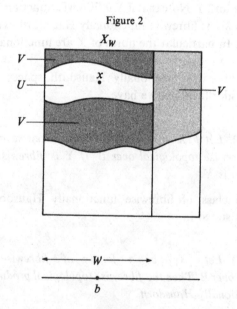

† The definition of fibrewise regularity adopted here is, *a priori*, somewhat weaker than the corresponding definition in [32].

For let $x \in X_b$, where $b \in B$, and let V be a neighbourhood of x in X. Then $V = \phi^{-1}V'$ where V' is a neighbourhood of $x' = \phi(x)$ in X'. If X' is fibrewise regular there exists a neighbourhood W of b in B and a neighbourhood U' of x' in X'_W such that $X'_W \cap \bar{U}' \subset V'$. Then $U = \phi^{-1}U'$ is a neighbourhood of x in X_W such that $X_W \cap \bar{U} \subset V$, as required.

Proposition (2.17). *Let $\{X_r\}$ be a finite family of fibrewise regular spaces over B. Then the fibrewise topological product $X = \prod_B X_r$ is fibrewise regular.*

In other words the class of fibrewise regular spaces is fibrewise multiplicative. For let $x \in X_b$, where $b \in B$. Consider a neighbourhood $V = \prod_B V_r$ of x in X, where V_r is a neighbourhood of $\pi_r(x) = x_r$ in X_r for each index r. Since X_r is fibrewise regular there exists a neighbourhood W_r of b in B and a neighbourhood U_r of x_r in $X_r | W_r$ such that the closure $(X_r | W_r) \cap \bar{U}_r$ of U_r in $X_r | W_r$ is contained in V_r. Then the intersection W of the W_r is a neighbourhood of b and $U = \prod_B U_r$ is a neighbourhood of x in X_W such that the closure $X_W \cap \bar{U}$ of U in X_W is contained in V as required. The same conclusion holds for infinite fibrewise products provided each of the factors is fibrewise non-empty.

Proposition (2.18). *Let $\phi : X \to Y$ be an open, closed and continuous fibrewise surjection, where X and Y are fibrewise topological over B. If X is fibrewise regular then so is Y.*

For let $y \in Y_b$, where $b \in B$, and let V be a neighbourhood of y. Pick $x \in \phi^{-1}(y)$. Then $U = \phi^{-1}V$ is a neighbourhood of x. If X is fibrewise regular there exists a neighbourhood W of b and a neighbourhood U' of x such that $X_W \cap \bar{U}' \subset U$. Then

$$Y_W \cap \phi(\bar{U}') \subset \phi U = V.$$

But $\phi \bar{U}' = \overline{\phi U'}$, since ϕ is closed, and $\phi U'$ is a neighbourhood of y since ϕ is open. Thus Y is fibrewise regular, as asserted.

Proposition (2.19). *Let X be fibrewise regular and fibrewise T_0 over B. Then X is fibrewise Hausdorff.*

For let $x, x' \in X_b$, where $b \in B$ and $x \neq x'$. Renaming the points, if necessary, we may suppose that x admits a neighbourhood V which does not contain x'. Since X is fibrewise regular there exists a neighbourhood W of b in B and a neighbourhood U of x in X_W such that the closure $X_W \cap \bar{U}$ of

U in X_W is contained in V. Then U and $X_W - X_W \cap \bar{U}$ are disjoint neighbourhoods of x and x', respectively, as required. Hence and from (2.18) we obtain

Proposition (2.20). *Let $\phi: X \to Y$ be an open, closed and continuous fibrewise surjection, where X and Y are fibrewise topological over B. Suppose that X is fibrewise regular and fibrewise T_0. Then Y is fibrewise Hausdorff.*

The functional version of the fibrewise regularity condition is stronger than the non-functional version but its properties are fairly similar. In the ordinary theory the term completely regular is always used instead of functionally regular and we extend this usage to the fibrewise theory.

Definition (2.21). *The fibrewise topological space X over B is fibrewise completely regular if for each point $x \in X_b$, where $b \in B$, and for each neighbourhood V of x there exists a neighbourhood W of b and a continuous function $\alpha: X_W \to I$ such that $\alpha(x) = 1$ and such that $\alpha = 0$ away from V.*

When the neighbourhoods of x are given by a fibrewise basis it is sufficient if the condition is satisfied for all fibrewise basic neighbourhoods. For example $B \times T$ is fibrewise completely regular for all completely regular T. Note that if X is fibrewise completely regular over B then $X_{B'}$ is fibrewise completely regular over B' for each subspace B' of B.

Subspaces of fibrewise completely regular spaces are fibrewise completely regular. In fact we have

Proposition (2.22). *Let $\phi: X \to X'$ be a fibrewise embedding, where X and X' are fibrewise topological over B. If X' is fibrewise completely regular then so is X.*

The proof is similar to that of (2.16) and will be omitted. We shall, however, prove that fibrewise completely regular spaces are finitely multiplicative as stated in

Proposition (2.23). *Let $\{X_r\}$ be a finite family of fibrewise completely regular spaces over B. Then the fibrewise topological product $X = \prod_B X_r$ is fibrewise completely regular.*

For let $x \in X_b$, where $b \in B$. Consider a fibrewise basic neighbourhood $\prod_B V_r$ of x in X, where V_r is a neighbourhood of $\pi_r(x) = x_r$ in X_r for each index r. Since X_r is fibrewise completely regular there exists a neighbourhood W_r of b and a continuous function $\alpha_r \colon X_{W_r} \to I$ such that $\alpha_r(x_r) = 1$ and such that $\alpha_r = 0$ away from V_r. Then the intersection W of the W_r is a neighbourhood of b and $\alpha \colon X_W \to I$ is a continuous function where

$$\alpha(\xi) = \inf_{r = 1,\ldots,n} \{\alpha_r(\xi_r)\}$$

for $\xi = (\xi_r) \in X_W$. Since $\alpha(x) = 1$ and since $\alpha = 0$ away from $X_W \cap \prod_B V_r$, this proves the result. The same conclusion holds for infinite fibrewise products provided that each of the factors is fibrewise non-empty.

Proposition (2.24). *Let* $\phi \colon X \to Y$ *be an open, closed and continuous fibrewise surjection, where* X *and* Y *are fibrewise topological over* B. *If* X *is fibrewise completely regular then so is* Y.

For let $y \in Y_b$, where $b \in B$, and let V be a neighbourhood of y. Pick $x \in X_b$, so that $U = \phi^{-1} V$ is a neighbourhood of x. Since X is fibrewise completely regular there exists a neighbourhood W of b and continuous function $\alpha \colon X_W \to I$ such that $\alpha(x) = 1$ and such that $\alpha = 0$ away from U. Using (1.3) we obtain a continuous function $\beta \colon Y_W \to I$ such that $\beta(y) = 1$ and such that $\beta = 0$ away from V.

Proposition (2.25). *Let* X *be fibrewise completely regular and fibrewise* T_0 *over* B. *Then* X *is fibrewise functionally Hausdorff.*

The proof is similar to that of (2.19) and will be omitted. Hence and from the previous result we obtain

Proposition (2.26). *Let* $\phi \colon X \to Y$ *be an open, closed and continuous fibrewise surjection, where* X *and* Y *are fibrewise topological spaces over* B. *If* X *is fibrewise completely regular and fibrewise* T_0 *then* Y *is fibrewise functionally Hausdorff.*

Definition (2.27). *The fibrewise topological space* X *over* B *is fibrewise normal if for each point* b *of* B *and each pair* H, K *of disjoint closed sets of* X, *there exists a neighbourhood* W *of* b *and a pair* U, V *of disjoint neighbourhoods of* $X_W \cap H$, $X_W \cap K$ *in* X_W (*see Figure* 3).

If X is fibrewise normal over B then clearly $X_{B'}$ is fibrewise normal over B' for each subspace B' of B. It is not difficult to see that closed subspaces of fibrewise normal spaces are fibrewise normal. In fact we have

Proposition (2.28). *Let $\phi: X \to X'$ be a closed fibrewise embedding, where X and X' are fibrewise topological over B. If X' is fibrewise normal then so is X.*

For let b be a point of B and let H, K be disjoint closed sets of X. Then $\phi H, \phi K$ are disjoint closed sets of X'. If X' is fibrewise normal there exists a neighbourhood W of b and disjoint neighbourhoods U, V of $X'_W \cap \phi H$, $X'_W \cap \phi K$ in X'_W. Then $\phi^{-1}U$, $\phi^{-1}V$ are disjoint neighbourhoods of $X_W \cap H$, $X_W \cap K$ in X_W.

Proposition (2.29). *Let $\phi: X \to Y$ be a closed continuous fibrewise surjection, where X and Y are fibrewise topological over B. If X is fibrewise normal then so is Y.*

For let b be a point of B and let H, K be disjoint closed sets of Y. Then $\phi^{-1}H$, $\phi^{-1}K$ are disjoint closed sets of X. If X is fibrewise normal there exists a neighbourhood W of b and disjoint open neighbourhoods U, V of $X_W \cap \phi^{-1}H$, $X_W \cap \phi^{-1}K$. Since ϕ is closed the sets

$$Y_W - \phi(X_W - U), \qquad Y_W - \phi(X_W - V)$$

Figure 3

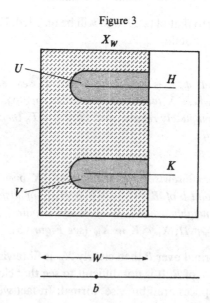

are open in Y_W, and form a disjoint pair of neighbourhoods of $Y_W \cap H$, $Y_W \cap K$ in Y_W as required.

Definition (2.30). *The fibrewise topological space X over B is fibrewise functionally normal if for each neighbourhood W of b, where $b \in B$, and each pair H, K of disjoint closed sets of X there exists a neighbourhood W of b and a continuous function $\alpha\colon X_W \to I$ such that $\alpha = 0$ throughout H_W and $\alpha = 1$ throughout K_W.*

For example, $B \times T$ is fibrewise functionally normal whenever T is normal. If X is functionally normal over B then clearly $X_{B'}$ is fibrewise functionally normal over B' for each subspace B' of B.

It is not difficult to see that closed subspaces of fibrewise functionally normal spaces are fibrewise functionally normal. In fact we have

Proposition (2.31). *Let $\phi\colon X \to X'$ be a closed fibrewise embedding, where X and X' are fibrewise topological over B. If X' is fibrewise functionally normal then so is X.*

Let b be a point of B and let H, K be disjoint closed sets of X. Then ϕH, ϕK are disjoint closed sets of X'. If X' is fibrewise functionally normal there exists a neighbourhood W of b and a continuous function $\alpha\colon X'_W \to I$ such that $\alpha = 0$ throughout $X'_W \cap \phi H$ and $\alpha = 1$ throughout $X'_W \cap \phi K$. Then $\beta = \alpha \circ \phi$ is a continuous function $X_W \to I$ such that $\beta = 0$ throughout $X_W \cap H$ and $\beta = 1$ throughout $X_W \cap K$. This proves (2.31).

Proposition (2.32). *Let $\phi\colon X \to Y$ be an open and closed continuous fibrewise surjection, where X and Y are fibrewise topological over B. If X is fibrewise functionally normal then so is Y.*

For let b be a point of B and let H, K be disjoint closed sets of Y. Then $\phi^{-1}H$, $\phi^{-1}K$ are disjoint closed sets of X. If X is fibrewise functionally normal there exists a neighbourhood W of b and a continuous function $\alpha\colon X_W \to I$ such that $\alpha = 0$ throughout $X_W \cap \phi^{-1}H$ and $\alpha = 1$ throughout $X_W \cap \phi^{-1}K$. Now a function $\beta\colon Y_W \to I$ is given by

$$\beta(y) = \sup_{x \in \phi^{-1}(y)} \alpha(x) \qquad (y \in Y_W).$$

Since ϕ is open and closed, as well as continuous, it follows that β is continuous. Since $\beta = 0$ on $Y_W \cap H$ and $\beta = 1$ on $Y_W \cap K$ this proves (2.32).

3. Fibrewise compact spaces

Definition (3.1). *The fibrewise topological space X over B is fibrewise compact if the projection p is proper.*

A useful characterization of fibrewise compact spaces is given by

Proposition (3.2). *The fibrewise topological space X over B is fibrewise compact if and only if X is fibrewise closed and every fibre of X is compact.*

For example the topological product $B \times T$ is fibrewise compact, for all compact T. For another example, if G is a compact subgroup of the topological group Γ then Γ is fibrewise compact over the factor space Γ/G.

Another characterization we shall be using occasionally is given by

Proposition (3.3). *Let X be fibrewise topological over B. Then X is fibrewise compact if and only if for each fibre X_b of X and each covering Γ of X_b by open sets of X there exists a neighbourhood W of b such that a finite subfamily of Γ covers X_W.*

These are special cases of well-known results, as are (3.4)–(3.7) below.

Proposition (3.4). *Let $\phi: X \to Y$ be a proper fibrewise function, where X and Y are fibrewise topological over B. If Y is fibrewise compact then so is X.*

In particular this holds when ϕ is a closed fibrewise embedding; thus closed subspaces of fibrewise compact spaces are fibrewise compact. The class of fibrewise compact spaces is multiplicative, as stated in

Proposition (3.5). *Let $\{X_r\}$ be a family of fibrewise compact spaces over B. Then the fibrewise topological product $\prod_B X_r$ is fibrewise compact.*

The general case is a consequence of the fibrewise Tychonoff theorem (see (4.4)). For finite products a simple argument can be used. Thus let X and Y be fibrewise topological over B. If X is fibrewise compact then the projection

$$p \times \mathrm{id}: X \times_B Y \to B \times_B Y \equiv Y$$

is proper, by (1.4). If Y is also fibrewise compact then so is $X \times_B Y$, by (3.4). A similar result holds for finite coproducts.

Proposition (3.6). *Let X be fibrewise topological over B. Suppose that X_j is fibrewise compact for each member X_j of a finite covering of X. Then X is fibrewise compact.*

Proposition (3.7). *Let $\phi: X \to Y$ be a continuous fibrewise surjection, where X and Y are fibrewise topological over B. If X is fibrewise compact then so is Y.*

Proposition (3.8). *Let X be fibrewise compact over B. Then $X_{B'}$ is fibrewise compact over B' for each subset B' of B.*

Proposition (3.9). *Let X be fibrewise topological over B. Suppose that X_{B_j} is fibrewise compact over B_j for each member B_j of an open covering of B. Then X is fibrewise compact over B.*

In fact the last result also holds for locally finite closed coverings, instead of open coverings.

Recall that the *subgraph* of a real-valued function $\alpha: B \to \mathbb{R}$ is the subset E of $B \times \mathbb{R}$ given by

$$E = \{(b, t) \in B \times \mathbb{R}: \quad 0 \leqslant t \leqslant \alpha(b)\}.$$

Here the fibres are compact, in any case. When α is continuous then E is fibrewise closed and so fibrewise compact. To see this we use (1.8). Thus let U be a neighbourhood of E_b, where $b \in B$. By compactness and the product topology there exists a neighbourhood W of b and a neighbourhood V of $[0, \alpha(b)]$ in $\mathbb{R}$ such that $E_b \subset E_W \cap V \subset U$. Since α is continuous there exists a neighbourhood $W' \subset W$ of b such that $\alpha W' \subset V$, and then $E_{W'} \subset U$ as required. We use this property of the subgraph in the proof of our next result.

The most familiar examples of compact spaces are closed and bounded subspaces of the Euclidean spaces $\mathbb{R}^n$ ($n = 0, 1, \ldots$). The corresponding result* in the fibrewise theory is

Proposition (3.10). *Let X be a closed subspace of $B \times \mathbb{R}^n$ ($n \geqslant 0$), regarded as a fibrewise topological space over B under the first projection. Then X is fibrewise compact if X is fibrewise bounded in the sense that there exists a continuous function $\alpha: B \to \mathbb{R}$ such that X_b is bounded by $\alpha(b)$ for each point $b \in B$.*

---

* The above result is due to Professor Albrecht Dold (private communication). A number of other results about subspaces of $B \times \mathbb{R}^n$ may be found in [18] and [19].

Suppose that the condition in (3.10) is satisfied. Since the Euclidean norm $D: \mathbb{R}^n \to \mathbb{R}$, given by $D(x) = \|x\|$, is proper the product

$$\text{id} \times D: B \times \mathbb{R}^n \to B \times \mathbb{R}$$

is also proper. Consider the subgraph E of α in $B \times \mathbb{R}$. Since $\text{id} \times D$ is proper so is the function $(\text{id} \times D)^{-1}E \to E$ which $\text{id} \times D$ determines. Also E is fibrewise compact and so $(\text{id} \times D)^{-1}E$ is fibrewise compact. But X is closed in $B \times \mathbb{R}^n$ and so closed in $(\text{id} \times D)^{-1}E$. Therefore X is fibrewise compact, as asserted. For example, the subset $\{(t, x) \in \mathbb{R} \times \mathbb{R}^n : \|x\| \leqslant t\}$ is fibrewise compact over $\mathbb{R}$.

Proposition (3.11). *Let $\phi: X \to Y$ be a fibrewise function, where X and Y are fibrewise topological over B. If X is fibrewise compact and*

$$\text{id} \times \phi: X \times_B X \to X \times_B Y$$

is proper then ϕ is proper.

For consider the commutative diagram shown below

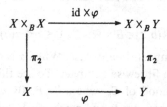

If X is fibrewise compact then π_2 is proper. If $\text{id} \times \phi$ is also proper then $\pi_2(\text{id} \times \phi) = \phi\pi_2$ is proper, and so ϕ itself is proper.

Definition (3.12). *The fibrewise topological space X over B is fibrewise locally compact† if for each point x of X_b, where $b \in B$, there exists a neighbourhood W of b and a neighbourhood $U \subset X_W$ of x such that the closure $X_W \cap \bar{U}$ of U in X_W is fibrewise compact over W.*

Clearly fibrewise compact spaces are necessarily fibrewise locally compact, likewise fibrewise regular fibrewise discrete spaces. Also the product space $B \times T$ is fibrewise locally compact for all locally compact T.

Closed subspaces of fibrewise locally compact spaces are fibrewise locally compact. In fact we have

---

† The definition of fibrewise local compactness adopted here is, *a priori*, somewhat weaker than the corresponding definition in [32].

Proposition (3.13). *Let* $\phi: X \to Y$ *be a closed fibrewise embedding, where* X *and* Y *are fibrewise topological over* B. *If* Y *is fibrewise locally compact then so is* X.

For let $x \in X_b$, where $b \in B$. If Y is fibrewise locally compact there exists a neighbourhood W of b and a neighbourhood $V \subset Y_W$ of $\phi(x)$ such that the closure $Y_W \cap \bar{V}$ of V in Y_W is fibrewise compact over W. Then $\phi^{-1}V \subset X_W$ is a neighbourhood of x such that the closure $X_W \cap \overline{\phi^{-1}V} = \phi^{-1}(Y_W \cap \bar{V})$ of $\phi^{-1}V$ in X_W is fibrewise compact over W.

Proposition (3.14). *Let* X *be fibrewise locally compact and fibrewise regular over* B. *Then for each point* x *of* X_b, *where* $b \in B$, *and each neighbourhood* V *of* x *in* X, *there exists a neighbourhood* U *of* x *in* X_W *such that the closure* $X_W \cap \bar{U}$ *of* U *in* X_W *is fibrewise compact over* W *and contained in* V.

Since X is fibrewise locally compact there exists a neighbourhood W' of b in B and a neighbourhood U' of x in $X_{W'}$ such that the closure $X_{W'} \cap \bar{U}'$ of U' in $X_{W'}$ is fibrewise compact over W'. Since X is fibrewise regular there exists a neighbourhood $W \subset W'$ of b and a neighbourhood U of x in X_W such that the closure $X_W \cap \bar{U}$ of U in X_W is contained in $X_W \cap U' \cap V$. Now $X_W \cap \bar{U}'$ is fibrewise compact over W, since $X_{W'} \cap \bar{U}'$ is fibrewise compact over W', and $X_W \cap \bar{U}$ is closed in $X_W \cap \bar{U}'$. Hence $X_W \cap \bar{U}$ is fibrewise compact over W and contained in V as required.

Proposition (3.15). *Let* $\phi: X \to Y$ *be an open continuous fibrewise surjection, where* X *and* Y *are fibrewise topological over* B. *If* X *is fibrewise locally compact and fibrewise regular then so is* Y.

For let y be a point of Y_b, where $b \in B$, and let V be a neighbourhood of y in Y. Pick any point x of $\phi^{-1}(y)$; then $\phi^{-1}V$ is a neighbourhood of x in X. If X is fibrewise locally compact there exists a neighbourhood W of b in B and a neighbourhood U of x in X_W such that the closure $X_W \cap \bar{U}$ of x in X_W is fibrewise compact over W and is contained in $\phi^{-1}V$. Then ϕU is a neighbourhood of y in Y_W, since ϕ is open, and the closure $Y_W \cap \overline{\phi U}$ of ϕU in Y_W is fibrewise compact over W and contained in V, as required.

Proposition (3.16). *Let* $\{X_r\}$ *be a finite family of fibrewise locally compact spaces over* B. *Then the fibrewise topological product* $\prod_B X_r$ *is fibrewise locally compact.*

In other words, the class of fibrewise locally compact spaces is finitely multiplicative. The proof is similar to that of (3.5.).

Proposition (3.17). *Let X be fibrewise locally compact and fibrewise regular over B. Let C be a compact subset of X_b, where $b \in B$, and let V be a neighbourhood of C in X. Then there exists a neighbourhood W of b in B and a neighbourhood U of C in X_W such that the closure $X_W \cap \bar{U}$ of U in X_W is fibrewise compact over W and contained in V.*

Since X is fibrewise locally compact there exists for each point x of C a neighbourhood W_x of b in B and a neighbourhood U_x of x in X_{W_x} such that the closure $X_{W_x} \cap \bar{U}_x$ of U_x in X_{W_x} is fibrewise compact over W_x and contained in V. The family $\{U_x : x \in C\}$ constitutes a covering of the compact C by open sets of X. Extract a finite subcovering indexed by $x_1, \ldots, x_n$, say. Take W to be the intersection

$$W_{x_1} \cap \cdots \cap W_{x_n},$$

and take U to be the restriction to X_W of the union

$$U_{x_1} \cup \cdots \cup U_{x_n}.$$

Then W is a neighbourhood of b in B and U is a neighbourhood of C in X_W such that the closure $X_W \cap \bar{U}$ of U in X_W is fibrewise compact over W and contained in V, as required.

Proposition (3.18). *Let $\phi : X \to Y$ be a proper fibrewise surjection, where X and Y are fibrewise topological over B. If X is fibrewise locally compact and fibrewise regular then so is Y.*

For let $y \in Y_b$, where $b \in B$, and let V be a neighbourhood of y in Y. Then $\phi^{-1}V$ is a neighbourhood of $\phi^{-1}(y)$ in X. Suppose that X is fibrewise locally compact. Since $\phi^{-1}(y)$ is compact there exists, by (3.17), a neighbourhood W of b in B and a neighbourhood U of $\phi^{-1}(y)$ in X_W such that the closure $X_W \cap \bar{U}$ of U in X_W is fibrewise compact over W and contained in $\phi^{-1}V$. Since ϕ is closed there exists a neighbourhood U' of y in Y_W such that $\phi^{-1}U' \subset U$. Then the closure $Y_W \cap \bar{U}'$ of U' in Y_W is contained in $\phi(X_W \cap \bar{U})$ and so is fibrewise compact over W. Since $Y_W \cap \bar{U}'$ is contained in V this shows that Y is fibrewise locally compact, as asserted.

We now come to a series of results in which fibrewise compactness (or fibrewise local compactness in some cases) is required as well as one of the fibrewise separation conditions discussed in the previous section.

Proposition (3.19). *Let $\phi: X \to Y$ be a continuous fibrewise function, where X is fibrewise compact and Y is fibrewise Hausdorff over B. Then ϕ is proper.*

For consider the diagram shown below, where r is the standard fibrewise topological equivalence and Γ is the fibrewise graph of ϕ.

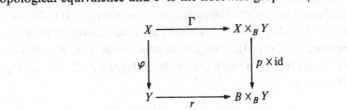

Now Γ is a closed embedding, by (2.10), since Y is fibrewise Hausdorff. Thus Γ is proper. Also p is proper and so $p \times \mathrm{id}$ is proper. Hence $(p \times \mathrm{id})\Gamma = r\phi$ is proper and so ϕ is proper, since r is a fibrewise topological equivalence.

Corollary (3.20). *Let $\phi: X \to Y$ be a continuous fibrewise injection, where X is fibrewise compact and Y is fibrewise Hausdorff. Then ϕ is a closed embedding.*

The corollary is often used in the case when ϕ is surjective to show that ϕ is a fibrewise topological equivalence.

Proposition (3.21). *Let $\phi: X \to Y$ be a proper fibrewise surjection, where X and Y are fibrewise topological over B. If X is fibrewise Hausdorff then so is Y.*

For since ϕ is a proper surjection so is $\phi \times \phi$, in the following diagram.

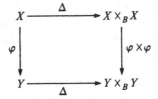

The diagonal ΔX is closed, since X is fibrewise Hausdorff, hence $(\phi \times \phi)\Delta X = \Delta \phi X$ is closed. But $\Delta \phi X = \Delta Y$, since ϕ is surjective, and so Y is fibrewise Hausdorff, as asserted.

Proposition (3.22). *Let X be fibrewise compact and fibrewise Hausdorff over B. Then X is fibrewise regular.*

For let $x \in X_b$, where $b \in B$, and let U be a neighbourhood of x in X. Since X is fibrewise Hausdorff there exists for each point $x' \in X_b$ such that $x' \notin U$ a neighbourhood $V_{x'}$ of x and a neighbourhood $V'_{x'}$ of x' which do not intersect. Now the family of open sets $V'_{x'}$, for $x' \in (X - U)_b$, forms a covering of $(X - U)_b$. Since $X - U$ is closed in X and therefore fibrewise compact there exists, by (3.3), a neighbourhood W of b in B such that $X_W - X_W \cap U$ is covered by a finite subfamily, indexed by $x'_1, \ldots, x'_n$, say. Now the intersection

$$V = V_{x_1} \cap \cdots \cap V_{x_n}$$

is a neighbourhood of x which does not meet the neighbourhood

$$V' = V'_{x'_1} \cup \cdots \cup V'_{x'_n}$$

of $X_W - X_W \cap U$. Therefore the closure $X_W \cap \bar{V}$ of $X_W \cap V$ in X_W is contained in U as required. An alternative proof is given in the next section.

We extend this last result to

Proposition (3.23). *Let X be fibrewise locally compact and fibrewise Hausdorff over B. Then X is fibrewise regular.*

For let $x \in X_b$, where $b \in B$, and let V be a neighbourhood of x in X. Let W be a neighbourhood of b in B and let U be a neighbourhood of x in X_W such that the closure $X_W \cap \bar{U}$ of U in X_W is fibrewise compact over W. Then $X_W \cap \bar{U}$ is fibrewise regular over W, by (3.22), since $X_W \cap \bar{U}$ is fibrewise Hausdorff over W. So there exists a neighbourhood $W' \subset W$ of b in B and a neighbourhood U' of x in $X_{W'}$ such that the closure $X_{W'} \cap \bar{U}'$ of U' in $X_{W'}$ is contained in $U \cap V \subset V$, as required.

Proposition (3.24). *Let X be fibrewise regular over B and let K be a fibrewise compact subset of X. Let b be a point of B and let V be a neighbourhood of K_b in X. Then there exists a neighbourhood W of b in B and a neighbourhood U of K_W in X_W such that the closure $X_W \cap \bar{U}$ of U in X_W is contained in V.*

We may suppose that K_b is non-empty since otherwise we can take $U = X_W$, where $W = B - p(X - V)$. Since V is a neighbourhood of each point x of K_b there exists, by fibrewise regularity, a neighbourhood W_x of b and a neighbourhood $U_x \subset X_{W_x}$ of x such that the closure $X_{W_x} \cap \bar{U}_x$ of U_x in X_{W_x} is contained in V. The family of open sets $\{X_{W_x} \cap U_x : x \in K_b\}$

covers K_b and so there exists a neighbourhood W' of b and a finite subfamily indexed by $x_1, \ldots, x_n$, say, which covers K_W. Then the conditions are satisfied with

$$W = W' \cap W_{x_1} \cap \cdots \cap W_{x_n}, \qquad U = U_{x_1} \cup \cdots \cup U_{x_n}.$$

Corollary (3.25). *Let X be fibrewise compact and fibrewise regular over B. Then X is fibrewise normal.*

Proposition (3.26). *Let X be fibrewise regular over B and let K be a fibrewise compact subset of X. Let $\{V_j\}$ $(j = 1, \ldots, n)$ be a covering of K_b $(b \in B)$ by open sets of X. Then there exists a neighbourhood W of b and a covering $\{U_j\}$ $(j = 1, \ldots, n)$ of K_W by open sets of X_W such that the closure $X_W \cap \bar{U}_j$ of U_j in X_W is contained in V_j for each j.*

To see this, write $V = V_2 \cup \cdots \cup V_n$, so that $X - V$ is closed in X. Hence $K \cap (X - V)$ is closed in K and so fibrewise compact. Applying the previous result to the neighbourhood V_1 of $K_b \cap (X - V)_b$ we obtain a neighbourhood W of b and a neighbourhood U of $K_W \cap (X - V)_W$ such that $X_W \cap \bar{U} \subset V_1$. Now $K \cap V$ and $K \cap (X - V)$ cover K, hence V and U cover K_W. Thus $U = U_1$ is the first step in the shrinking process. We continue by repeating the argument for $\{U_1, V_2, \ldots, V_n\}$, so as to shrink V_2, and so on. Hence the result is obtained.

Proposition (3.27). *Let $\phi: X \to Y$ be a proper fibrewise surjection, where X and Y are fibrewise topological over B. If X is fibrewise regular then so is Y.*

For let X be fibrewise regular. Let y be a point of Y_b $(b \in B)$ and let V be a neighbourhood of y in Y. Then $\phi^{-1}V$ is a neighbourhood of the compact $\phi^{-1}(y)$ in X. By (3.24), therefore, there exists a neighbourhood W of b in B and a neighbourhood U of $\phi^{-1}(y)$ in X_W such that the closure $X_W \cap \bar{U}$ of U in X_W is contained in $\phi^{-1}V$. Now since ϕ_W is closed there exists a neighbourhood V' of y in Y_W such that $\phi^{-1}V' \subset U$, and then the closure $X_W \cap \bar{V}'$ of V' in X_W is contained in V since

$$\mathrm{Cl}\, V' = \mathrm{Cl}(\phi\phi^{-1}V') = \phi\,\mathrm{Cl}(\phi^{-1}V') \subset \phi\,\mathrm{Cl}\, U \subset \phi\phi^{-1}V \subset V.$$

Thus Y is fibrewise regular as asserted.

4. Tied filters

In all but the most elementary expositions of topology the concept of filter nowadays plays a central role, particularly where compactness is involved. The corresponding role in fibrewise topology is played by the concept of tied filter, as follows. Let us return to the category of fibrewise sets over a topological space, since this is where the concept of tied filter originates. Thus let B be a topological space and let X be a fibrewise set over B with projection p. By a *tied filter* on X I mean a pair $(b, \mathscr{F})$, where b is a point of B and $\mathscr{F}$ is a filter on X such that b is a limit point of the filter $p_* \mathscr{F}$ on B. In this situation it is usually sufficient just to say that $\mathscr{F}$ is tied to b or that $\mathscr{F}$ is a b-filter.

Given a b-filter $\mathscr{F}$ on X I describe a family Γ of members of $\mathscr{F}$ as a *fibrewise basis* for $\mathscr{F}$ if for each member M of $\mathscr{F}$ we have $X_W \cap E \subset M$ for some member E of Γ and some neighbourhood W of b. In this situation I also say that Γ *generates* $\mathscr{F}$. Thus although Γ as it stands may not serve as a basis for $\mathscr{F}$ in the ordinary sense, a basis can always be obtained by taking the intersections $X_W \cap M$ where $M \in \Gamma$ and W is a neighbourhood of b. A *fibrewise subbasis* is a family which forms a fibrewise basis after finite intersections have been taken.

Changing our viewpoint, suppose that we have a family Γ of subsets of X such that (i) $X_W \cap E$ is non-empty for each member E of Γ and neighbourhood W of b and such that (ii) if $E_i \in \Gamma$ $(i = 1, 2)$ then $X_W \cap F \subset E_1 \cap E_2$ for some member F of Γ and some neighbourhood W of b. Then a filter $\mathscr{F}$ on X is generated by the intersections $X_W \cap E$, where E runs through the members of Γ and W runs through the neighbourhoods of b. Clearly $\mathscr{F}$ is a b-filter and Γ is a fibrewise basis for $\mathscr{F}$ in the previous sense.

A refinement (in the ordinary sense) of a b-filter is necessarily a b-filter. Hence a b-filter which is an ultrafilter (in the ordinary sense) is a b-ultrafilter, i.e. maximal in the class of b-filters. We recall (see [9])

Proposition (4.1). *Let $\mathscr{F}$ be a b-ultrafilter $(b \in B)$ on the fibrewise set X over B. Let $\{X_r\}$ be a finite family of subsets of X of which the union belongs to $\mathscr{F}$. Then X_r belongs to $\mathscr{F}$ for some index r.*

Note that if $\phi \colon X \to Y$ is a fibrewise function, where X and Y are fibrewise sets over B, then the direct images of the members of a b-filter $\mathscr{F}$ on X $(b \in B)$ generate a b-filter $\phi_* \mathscr{F}$ on Y; moreover $\phi_* \mathscr{F}$ is a b-ultrafilter if $\mathscr{F}$ is a b-ultrafilter. Furthermore, if $\mathscr{G}$ is a b-filter on Y, and if the trace of $\mathscr{G}$ on ϕX is defined (as is always the case when ϕ is surjective) then the inverse images of the members of $\mathscr{G}$ generate a b-filter $\phi^* \mathscr{G}$ on X.

We now return to the situation where X is fibrewise topological over B. By an *adherence point* of a b-filter $\mathscr{F}$ $(b \in B)$ on X I mean a point of the fibre X_b which is an adherence point of $\mathscr{F}$ as a filter on the set X. Points outside X_b are not to be regarded as adherent. The term *limit point* will be used similarly; for example if $\mathscr{N}_x$ is the neighbourhood filter in X of a point $x \in X_b$ then x is a limit point of $\mathscr{N}_x$ as a b-filter. Of course a limit point is necessarily an adherence point. It must be emphasized that a b-filter $\mathscr{F}$ is not to be regarded as convergent if $\mathscr{F}$, as a filter, converges to a point outside X_b. Observe that if $\mathscr{F}'$ is a refinement of $\mathscr{F}$, where $\mathscr{F}$ and $\mathscr{F}'$ are b-filters on X, then an adherence point of $\mathscr{F}'$ is an adherence point of $\mathscr{F}$, while a limit point of $\mathscr{F}$ is a limit point of $\mathscr{F}'$. We have at once

Proposition (4.2). *Let $\phi: X \to Y$ be a fibrewise function, where X and Y are fibrewise topological over B. Then ϕ is continuous if and only if the following condition is satisfied: whenever $\mathscr{F}$ is a convergent tied filter on X with limit point x then $\phi_* \mathscr{F}$ is a convergent tied filter on Y with limit point $\phi(x)$.*

Also we recall

Proposition (4.3). *Let $\phi: X \to Y$ be a fibrewise function, where X and Y are fibrewise topological over B. Then ϕ is proper if and only if for each point $b \in B$, each b-filter $\mathscr{F}$ on X and each adherence point y of $\phi_* \mathscr{F}$ in Y there exists an adherence point $x \in \phi^{-1}(y)$ of $\mathscr{F}$ in X. Equivalently ϕ is proper if and only if for each b-ultrafilter $\mathscr{F}$ on X and each limit point y of $\phi_* \mathscr{F}$ in Y there exists a limit point $x \in \phi^{-1}(y)$ of $\mathscr{F}$ in X.*

Proposition (4.4). *Let $\{\phi_r\}$ be a family of proper fibrewise functions, where $\phi_r: X_r \to Y_r$, and X_r, Y_r are fibrewise topological over B. Then the fibrewise product*

$$\prod_B \phi_r: \prod_B X_r \to \prod_B Y_r$$

is also proper.

This result, which may be referred to as the *fibrewise Tychonoff theorem*, is best proved by using tied ultrafilters. Thus let $\mathscr{F}$ be a b-ultrafilter on $X = \prod_B X_r$, where $b \in B$. Then $\phi_* \mathscr{F}$ is a b-ultrafilter on $Y = \prod_B Y_r$, where $\phi = \prod_B \phi_r$. Suppose that $\phi_* \mathscr{F}$ converges to a limit point $y \in Y$. Then the projection $\rho_{r*} \phi_* \mathscr{F}$ converges to $y_r = \rho_r(y) \in Y_r$ for each index r. But $\rho_{r*} \phi_* = \phi_{r*} \pi_{r*}$, and so $\pi_{r*} \mathscr{F}$ converges to a limit point $x_r \in \phi_r^{-1}(y_r)$, since ϕ_r is proper. Then $\mathscr{F}$ itself converges to $x = (x_r) \in \phi^{-1}(y)$, and we obtain (4.4).

In a fibrewise topological space, as we have seen, the neighbourhood filter of a point is necessarily a tied filter. So if we wish to assign a fibrewise topology to a fibrewise set by specifying the neighbourhood filters these must, of course, be tied filters. Specifically, let X be a fibrewise set over B. Suppose that for each point b of B and each point x of X_b we have a b-filter $\mathcal{N}_x$, of which the principal b-filter $\mathscr{E}_x$ is a refinement. Suppose that the collection of tied filters satisfies the usual coherence condition: for each member M of $\mathcal{N}_x$ there exists a member N of $\mathcal{N}_x$ such that $M \in \mathcal{N}_\xi$ whenever $\xi \in N$. Then a fibrewise topology on X is defined so that $\mathcal{N}_x$ is the neighbourhood filter of x, for each $x \in X_b$, $b \in B$. The collection of tied filters is called the *fibrewise neighbourhood system* of X.

Proposition (4.5). *The fibrewise topological space X over B is fibrewise Hausdorff if and only if each convergent tied filter on X has a unique adherence point.*

In particular, limits of tied filters are unique in a fibrewise Hausdorff space. To prove (4.5) first suppose that the uniqueness condition is satisfied. Let $x, x' \in X_b$, where $b \in B$ and $x \neq x'$. The neighbourhood filter $\mathcal{N}_x$ of x is a convergent b-filter, and so x' is not an adherence point, by uniqueness. Hence there exist disjoint neighbourhoods of x, x', as required. Conversely suppose that X is fibrewise Hausdorff. Let $\mathscr{F}$ be a b-filter on X with limit point x, say. If $x' \in X_b$ is distinct from x there exist neighbourhoods V, V' of x, x' in X which are disjoint. But $V \in \mathscr{F}$, since $\mathscr{F}$ converges to x, and so x' cannot be an adherence point of $\mathscr{F}$.

Next we establish a result which will play an important role in Chapter III.

Proposition (4.6). *Let X be fibrewise topological over B, and let A be a dense subspace of X. Let $\phi: A \to Y$ be a continuous fibrewise function, where Y is fibrewise regular and fibrewise Hausdorff. Suppose that for each point b of B, if a b-filter $\mathscr{F}$ on A converges in X, then the corresponding b-filter $\phi_*\mathscr{F}$ on Y converges in Y. Then there exists one and only one continuous fibrewise function $\psi: X \to Y$ such that $\psi \mid A = \phi$.*

To define ψ at a given point $x \in X_b$ consider the neighbourhood b-filter $\mathcal{N}_x$ of x in X. Since A is dense this induces a b-filter $\mathcal{M}_x$ on A. Then the b-filter $\phi_*\mathcal{M}_x$ converges in Y, since $\mathcal{M}_x$ converges in X, and the limit point of $\phi_*\mathcal{M}_x$ is unique, by (4.5), since Y is fibrewise Hausdorff. We define $\psi(x)$ to be the limit point of $\phi_*\mathcal{M}_x$; clearly ψ is a fibrewise function.

To show that ψ is continuous we proceed as in the ordinary version of this result (see X.5.3 of [20], for example) and show that for each point x of X_b and each closed neighbourhood N of $\psi(x)$ in Y_W the inverse image $\psi^{-1}N$ is a neighbourhood of x in X. Now let U be an open neighbourhood of $\psi(x)$. Since Y is fibrewise regular there exists an open neighbourhood W of b and an open neighbourhood V of $\psi(x)$ such that the closure $Y_W \cap \bar{V}$ of V in Y_W is contained in U. Then

$$\psi^{-1}(Y_W \cap \bar{V}) = X_W \cap (\psi^{-1}\bar{V})$$

is a neighbourhood of x contained in $\psi^{-1}U$. Therefore ψ is continuous. Finally uniqueness follows at once from (4.5) since Y is fibrewise Hausdorff. We refer to ψ as the *continuous fibrewise extension* of ϕ.

Proposition (4.7). *Let X be fibrewise compact over B. Let $\mathcal{F}$ be a b-filter on X, where $b \in B$, and let $A \subset X_b$ be the adherence set of $\mathcal{F}$. Then every neighbourhood of A belongs to $\mathcal{F}$.*

For let V be a neighbourhood of A. Suppose, to obtain a contradiction, that every member of $\mathcal{F}$ meets the complement CV of V. The traces on CV of the members of $\mathcal{F}$ generate a b-filter $\mathcal{G}$ on X. Since X is fibrewise compact there exists an adherence point $\xi \in X_b$ of $\mathcal{G}$. Now $\xi \notin A$ since the neighbourhood V of A does not meet the member CV of $\mathcal{G}$. However ξ is an adherence point of $\mathcal{F}$, since $\mathcal{G}$ is a refinement of $\mathcal{F}$, and so we have our contradiction.

We can use (4.7) to give an alternative proof of (3.22), where X is fibrewise Hausdorff as well as fibrewise compact. Let $x \in X_b$, where $b \in B$, and let $\mathcal{F}$ be the b-filter generated by the closed neighbourhoods of x in X. Then x is the sole adherence point of $\mathcal{F}$ and so x is a limit point of $\mathcal{F}$, by (4.7). Therefore X is fibrewise regular, as asserted in (3.22).

5. Fibrewise quotient spaces

Let $\phi: X \to Y$ be a fibrewise surjection where X is a fibrewise topological space and Y is a fibrewise set over B. If we give Y the quotient topology, in the usual sense, then Y obtains the finest fibrewise topology which makes ϕ continuous. In that case we describe Y as a *fibrewise quotient space* and ϕ as a *fibrewise quotient map*. For example, ϕ is necessarily a fibrewise quotient map if ϕ is continuous and either open or closed. Note that if $\phi: X \to Y$ is a fibrewise quotient map over B then the restriction $\phi_{B'}: X_{B'} \to Y_{B'}$ is a fibrewise quotient map over B' when B' is an open or closed subspace of B. Fibrewise quotient maps can be characterized as follows.

Proposition (5.1). *Let $\phi: X \to Y$ be a fibrewise quotient map, where X and Y are fibrewise topological over B. Then for each fibrewise topological Z, a fibrewise function $\psi: Y \to Z$ is continuous if and only if the composition $\psi\phi: X \to Z$ is continuous.*

Corollary (5.2). *Let $\phi: X \to Y$ be a continuous fibrewise function, where X and Y are fibrewise topological spaces over B. If ϕ admits a right inverse then Y has the fibrewise quotient topology and ϕ is a fibrewise quotient map.*

The usual transitivity condition is satisfied. Specifically let $\phi: X \to Y$ and $\psi: Y \to Z$ be fibrewise surjections, where X is a fibrewise topological space and Y and Z are fibrewise sets over B. If first Y receives the fibrewise quotient topology from X, with respect to ϕ, and then Z receives the fibrewise quotient topology from Y, with respect to ψ, this is the same as if Z receives the fibrewise quotient topology directly from X with respect to $\psi\phi$.

Fibrewise products of fibrewise quotient maps are not necessarily fibrewise quotient maps. We prove

Proposition (5.3). *Let $\phi: X \to Y$ be a fibrewise quotient map, where X and Y are fibrewise topological over B. Then the fibrewise product*

$$\phi \times \mathrm{id}: X \times_B T \to Y \times_B T$$

is a fibrewise quotient map, for all fibrewise locally compact and fibrewise regular spaces T.

For let $U \subset X \times_B T$ be open and saturated with respect to $\psi = \phi \times \mathrm{id}$. We have to show that ψU is open in $Y \times_B T$. So let $(y, t) \in \psi U$, where $y \in Y_b$, $t \in T_b$, $b \in B$, and pick $x \in \phi^{-1}(y) \subset X_b$. We have $(x, t) \in U$, since U is saturated. Consider the subset N of T_b given by

$$\{x\} \times N = (\{x\} \times T_b) \cap U.$$

Now N is open in T_b, since U is open in $X \times_B T$, and so $N = M \cap T_b$ where M is open in T. Since T is fibrewise locally compact and fibrewise regular there exists, by (3.17), a neighbourhood W of b in B and a closed neighbourhood $K \subset M$ of t in T_W such that K is fibrewise compact over W. Consider the subset

$$V = \{\xi \in X_W: \{\xi\} \times_W K \subset U\}$$

of X_W. We have $(y, t) \in \phi V \times_W K \subset \psi U$. So to prove that ψU is a neighbourhood of (y, t) in $Y \times_B T$ it is sufficient to prove that ϕV is a neighbourhood of y in Y.

In fact V is open in X. For let $\xi \in V$ so that $\{\xi\} \times_W K_\beta \subset U$, where $\beta = p(\xi)$ and $p\colon X \to B$. Since K is fibrewise compact over W the projection

$$X_W \times_W K \to X_W \times_W W \to X_W$$

is closed. Since U is a neighbourhood of the inverse image $\{\xi\} \times K_\beta$ of ξ under the projection there exists a neighbourhood $W' \subset W$ of β and a neighbourhood V' of ξ in $X_{W'}$ such that $V' \times_{W'} K_{W'} \subset U$. This implies that $V' \subset V$, by definition of V, and so V is open.

Moreover V is saturated. For $V \subset \phi^{-1}\phi V$, as always. Also

$$\phi^{-1}\phi V \times_W K = \psi^{-1}\psi(V \times_W K) \subset \psi^{-1}\psi U = U.$$

Therefore $\phi^{-1}\phi V \subset V$, by definition of V, and so $\phi^{-1}\phi V = V$. Thus V is saturated, as well as open, and so ϕV is open. Since $y \in \phi V$ this completes the proof.

By a *fibrewise equivalence relation* on a fibrewise set X over B we mean an equivalence relation R, in the usual sense, such that each equivalence class is contained in a fibre of X. In other words, if $x \in X_b$ $(b \in B)$ then $\xi \in X_b$ whenever ξ is R-related to x. Thus R may be regarded as a subset of the fibrewise product $X \times_B X$, rather than the Cartesian product $X \times X$, and this is the point of view we shall be taking. The quotient set X/R of X with respect to R is again a fibrewise set and the natural projection $\pi\colon X \to X/R$ is a fibrewise function.

Of course a fibrewise function $\phi\colon X \to Y$, where X and Y are fibrewise sets over B, determines a fibrewise equivalence relation $R = \phi^{-2}\Delta Y$ on X, such that $\phi\pi^{-1}\colon X/R \to Y$ is injective. In particular the projection $p\colon X \to B$ determines an equivalence relation on X for which the equivalence classes are the fibres. I refer to this as the *trivial fibrewise equivalence relation*; in this case the fibrewise quotient set is equivalent to B itself.

Let us now turn to the situation where X is fibrewise topological over B. Given a fibrewise equivalence relation R on X we give X/R the fibrewise quotient topology, so as to make the natural projection π a quotient map.

For example, let X be fibrewise topological over B and let A be a closed subspace of X. Take R to be the fibrewise equivalence relation on $X +_B B$ such that the equivalence classes in the fibre $X_b + \{b\}$ $(b \in B)$ are (i) the individual points of $(X - A)_b$ and (ii) the set $A_b + \{b\}$. The fibrewise quotient space thus defined is called the *fibrewise collapse* of X with respect to A and denoted by $X/_B A$. Note that the image of B is closed in $X/_B A$. When A is empty the fibrewise collapse is just $X +_B B$, while $X/_B X$ reduces to B. Note that a continuous fibrewise function $\phi\colon X \to Y$, for any fibrewise topological Y, induces a continuous fibrewise function $\phi\pi^{-1}\colon X/_B A \to Y$ whenever ϕ is fibrewise constant on A.

In the case of a continuous fibrewise function $\phi\colon X \to Y$, where X and Y are fibrewise topological over B, the fibrewise function $\phi\pi^{-1}\colon X/R \to Y$ is an embedding, where R is the fibrewise equivalence relation associated with ϕ, provided ϕ is either (i) open or (ii) closed.

Proposition (5.4). *Let X be fibrewise topological over B and let A be a fibrewise compact closed subspace of X. Then the natural projection $\pi\colon X \to X/_B A$ to the fibrewise collapse is proper.*

Of course if X is fibrewise Hausdorff then the assumption that A is closed follows from the assumption that A is fibrewise compact. To show that π is closed observe that if $E \subset X$ the saturation $\pi^{-1}\pi E$ of E is just $E \cup A_{p(E \cap A)}$, where $p\colon X \to B$ is the projection. If E is closed in X then $E \cap A$ is closed in A, hence $p(E \cap A)$ is closed in B since $p \mid A$ is closed, hence $A_{p(E \cap A)}$ is closed in A and so closed in X. Therefore the saturation of a closed set is closed, and so π is closed.

We also need to show that the fibres of π are compact. Let $x \in X_b$, where $b \in B$. If $x \notin A_b$ then $\pi^{-1}\pi(x) = x$, which is compact, while if $x \in A_b$ then $\pi^{-1}\pi(x) = A_b$, which is compact since A is fibrewise compact. This completes the proof. By combining (5.4) with (3.18) and (3.21) we obtain

Corollary (5.5). *Let X be fibrewise Hausdorff over B and let A be a closed subspace of X. If A is fibrewise compact then the fibrewise collapse $X/_B A$ is fibrewise Hausdorff.*

Corollary (5.6). *Let X be fibrewise regular over B and let A be a closed subspace of X. If A is fibrewise compact then the fibrewise collapse $X/_B A$ is fibrewise regular.*

Definition (5.7). *Let $\phi\colon X \to Y$ be a fibrewise quotient map, where X and Y are fibrewise topological over B. Suppose that $\phi X' \subset Y'$, where $X' \subset X$ and $Y \subset Y$, and that ϕ maps $X - X'$ homeomorphically onto $Y - Y'$. Then ϕ is a fibrewise relative homeomorphism.*

In other words a fibrewise function is a fibrewise relative homeomorphism if it is a relative homeomorphism in the ordinary sense.

For example let X be fibrewise topological over B and let A be a closed subspace of X. Let R be a fibrewise equivalence relation on X such that $R \cap (X \times A) = R \cap (A \times A)$. Then the natural projection

$$(X, A) \to (X/R, A/R_A)$$

is a fibrewise relative homeomorphism, where R_A is the restriction of R to A.

If R is a fibrewise equivalence relation on the fibrewise topological space X there is not, in general, any very convenient necessary and sufficient condition which ensures that X/R is fibrewise Hausdorff. Of course it is necessary for the points of X/R to be closed or, equivalently, for the equivalence classes of R to be closed in X, but, in general, this is not sufficient.

Recall from (2.8) that the domain of a continuous fibrewise injection with fibrewise Hausdorff codomain is also fibrewise Hausdorff. Thus we obtain

Proposition (5.8). *Let $\phi: X \to Y$ be a continuous fibrewise function, where X is fibrewise topological and Y is fibrewise Hausdorff over B. If R is the fibrewise equivalence relation on X determined by ϕ then the fibrewise quotient space X/R is fibrewise Hausdorff.*

Proposition (5.9). *Let R be a fibrewise equivalence relation on the fibrewise Hausdorff space X over B. If there exists a section $s: X/R \to X$ then X/R is fibrewise Hausdorff and $s(X/R)$ is closed in X.*

In fact s is an embedding and so X/R is fibrewise topologically equivalent to $s(X/R)$ which is fibrewise Hausdorff. Moreover $s(X/R)$ is precisely the set of points x of X such that $s(\pi(x)) = x$, where $\pi: X \to X/R$ is the natural projection, and so $s(X/R)$ is closed, by (2.9).

This leads to the question of when $\pi: X \to X/R$ admits a section. Let us assume that X/R is fibrewise Hausdorff. Let us also assume that X contains a fibrewise compact subspace K which meets every equivalence class in precisely one point. Then the function $s: X/R \to X$ so defined is a section. For let $\rho = \pi \,|\, K$; then $X/R = \rho K$ is fibrewise compact, since ρ is continuous, and so closed, since X/R is fibrewise Hausdorff. Therefore the bijection $K \to X/R$ induced by ρ is a fibrewise topological equivalence.

Now suppose that we have a cotriad in our category of the form

$$X \overset{\phi}{\leftarrow} A \overset{\psi}{\to} Y,$$

where A, X and Y are fibrewise topological spaces over B, where ϕ is a fibrewise embedding and where ψ is a continuous fibrewise function. Then a fibrewise equivalence relation is defined on the fibrewise topological coproduct $X +_B Y$ in which $x \in X_b$ and $y \in Y_b$ ($b \in B$) are equivalent when $x = \phi(a)$ and $y = \psi(a)$ for some $a \in A_b$. The fibrewise quotient of $X +_B Y$ with respect to this relation is called the *fibrewise push-out* of the cotriad and denoted by $X +_B^A Y$. The fibrewise collapse is essentially the special case when $Y = B$.

In the fibrewise push-out the saturation of a subset X' of X is the union of X' with $\psi\phi^{-1}X'$ while the saturation of a subset Y' of Y is the union of Y' with $\phi\psi^{-1}Y'$. It follows that the projection

$$\pi: X +_B Y \to X +_B^A Y$$

is open when ϕ and ψ are open, and is closed when ϕ and ψ are closed. It also follows that π is proper when ϕ and ψ are proper. In that case $X +_B^A Y$ is fibrewise Hausdorff when X and Y are fibrewise Hausdorff, by (3.21), and $X +_B^A Y$ is fibrewise regular when X and Y are fibrewise regular, by (3.26).

To illustrate these results consider the following family of endofunctors of our category, which are of great importance in fibrewise homotopy theory, as we shall see in Chapter IV. Let T be fibrewise topological over B and let T_0 be a closed subspace of T. Define $\Phi_B(X)$, for any fibrewise topological X, to be the fibrewise push-out of the cotriad

$$T \times_B X \leftarrow T_0 \times_B X \to T_0,$$

and similarly for continuous fibrewise functions. In the case in which $(T, T_0) = B \times (T', T_0')$, where T' is a topological space and T_0' is a closed subspace of T', we may identify $\Phi_B(X)$ with the fibrewise push-out of the cotriad

$$T' \times X \leftarrow T_0' \times X \to T_0' \times B.$$

In particular when $(T', T_0') = (I, \{0\})$ the endofunctor Φ_B thus defined is known as the *fibrewise cone* and denoted by Γ_B, while when $(T', T_0') = (I, \{0, 1\})$ the endofunctor is known as the *fibrewise suspension* and denoted by Σ_B. The above results show that $\Phi_B(X)$ is fibrewise Hausdorff whenever X is fibrewise compact and fibrewise Hausdorff while $\Phi_B(X)$ is fibrewise regular whenever X is fibrewise compact and fibrewise regular. In particular these conclusions are true for Γ_B and Σ_B.

The maximal fibrewise Hausdorff quotient of a fibrewise regular space provides a further illustration of the properties of the fibrewise quotient topology, as follows. Given a fibrewise regular space X over B let us say that $\xi R \eta$, where $\xi, \eta \in X_b$, $b \in B$, if $\xi \in \overline{\{\eta\}}$. Clearly R is a fibrewise equivalence relation, moreover the projection $\pi: X \to X' = X/R$ is open and closed. For let U be an open set of X. If $\xi \in$ sat U, where $\xi \in X_b$, then $\xi R \eta$ for some $\eta \in U$, where $\eta \in X_b$. But $\eta \in \overline{\{\xi\}}$ and so $\xi \in U$. Thus open sets are saturated, therefore the projection is open. Again let H be a closed set of X. If $\xi \in \operatorname{sat} H$, where $\xi \in X_b$, then $\xi R \eta$ for some $\eta \in H$, where $\eta \in X_b$. But $\xi \in \overline{\{\eta\}} \subset \overline{H} = H$. Thus closed sets are also saturated, therefore the projection is closed as well as open. Applying (2.18) we obtain at once that X' is fibrewise regular.

In fact X' is also fibrewise Hausdorff. For suppose that ξ is unrelated to η, where $\xi, \eta \in X_b$, $b \in B$. Then $V = X - \overline{\{\eta\}}$ is a neighbourhood of ξ which does not contain η. By fibrewise regularity there exists a neighbourhood W of b and a neighbourhood U of ξ such that $X_W \cap \overline{U} \subset V$. Then $X_W \cap U$ and $X_W - X_W \cap \overline{U}$ are disjoint neighbourhoods of ξ and η, respectively. Since open sets are saturated, as we have seen, this proves the assertion.

Furthermore, if $\phi : X \to Y$ is a continuous fibrewise function, where Y is fibrewise Hausdorff, then ϕ can be factored through the continuous fibrewise function $\psi : X' \to Y$, where $\psi \pi = \phi$. For this reason X' may be called the *maximal fibrewise Hausdorff quotient* of X. We shall meet it again in Chapter III, where it will be denoted by $\tilde{X}_B$.

6. Fibrewise pointed topological spaces

Let us forget about the topology of the base space for a moment and return to the original situation as at the beginning of §1. By a *fibrewise pointed set* over B we mean a triple consisting of a set X together with functions

$$B \xrightarrow{s} X \xrightarrow{p} B$$

such that $ps = \mathrm{id}$. We refer to p as the *projection* and to s as the *section* of X. Usually X alone is sufficient notation. Note that a fibrewise pointed set X over B determines, by restriction, a fibrewise pointed set $X_{B'}$ over B' for each subset B' of B, with section $s_{B'}$ and projection $p_{B'}$. In particular, the fibre X_b over a point b of B can be regarded as a pointed set, with basepoint $s(b)$.

We regard B as a fibrewise pointed set over itself with the identity as section and projection. Moreover we regard $B \times T$ as a fibrewise pointed set over B, for each pointed set T, with section given by $b \to (b, t_0)$, where t_0 denotes the basepoint.

Fibrewise pointed sets form a category with the following notion of morphism. Let X, Y be fibrewise pointed sets over B with sections s, t

and projections p, q respectively. We describe a function $\phi: X \to Y$ as *fibrewise pointed* if $\phi s = t$ and $q\phi = p$, as shown below.

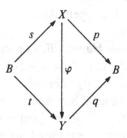

In other words ϕ is fibrewise pointed if ϕ is both fibre-preserving and section-preserving. The function $tp: X \to Y$ satisfies these conditions and is called the *fibrewise constant function*. Note that a fibrewise pointed function $\phi: X \to Y$ determines by restriction a fibrewise pointed function $\phi_{B'}: X_{B'} \to Y_{B'}$, for each subset B' of B. In particular ϕ determines a pointed function $\phi_b: X_b \to Y_b$ for each point b of B.

The equivalences in the category are called *fibrewise pointed equivalences*. Note that if ϕ is a fibrewise pointed equivalence over B then $\phi_{B'}$ is a fibrewise pointed equivalence over B' for each $B' \subset B$, moreover, ϕ_b is a pointed equivalence for each point b of B. Conversely if ϕ is a fibrewise pointed function such that ϕ_b is a pointed equivalence for each b then ϕ is a fibrewise pointed equivalence. If X is fibrewise pointed equivalent to $B \times T$ for some pointed set T we say that X is *trivial*, as a fibrewise pointed set over B.

Now consider the topological version of these concepts, with the topology restored to the base space. By a *fibrewise pointed topological space* over B we mean a triple consisting of a topological space X together with continuous functions

$$B \xrightarrow{s} X \xrightarrow{p} B$$

such that $ps = \mathrm{id}_B$. Thus X is a fibrewise topological space over B with section s. Usually X alone is sufficient notation. Note that the projection is necessarily a quotient map and the section is necessarily an embedding. Although we require the section to be a closed embedding in most of the results which follow, the restriction is often unnecessary when B is a point-space.

A fibrewise pointed topological space X over B determines, by restriction, a fibrewise pointed topological space $X_{B'}$ over B', for each subspace B' of B. In particular the fibre X_b over each point b of B can be regarded as a pointed topological space.

We regard B as a fibrewise pointed topological space over itself with the identity as section and projection. Moreover we regard $B \times T$ as a fibrewise pointed topological space over B, for each pointed topological space T, with section given by $b \to (b, t_0)$, where t_0 denotes the basepoint.

Fibrewise pointed topological spaces over a given base form a category using the continuous fibrewise pointed functions as morphisms. The equivalences in the category are called *fibrewise pointed topological equivalences*. Note that the fibrewise constant function, between fibrewise pointed topological spaces, is automatically continuous.

If X is fibrewise pointed topologically equivalent to $B \times T$ for some pointed topological space T we say that X is *trivial*, as a fibrewise pointed topological space over B. *Local triviality*, in the same sense, is defined similarly.

In the case of a fibrewise pointed topological space we restrict the term subspace to subspaces which contain the image of the section, so that the inclusion is a fibrewise pointed embedding.

Let X be fibrewise topological over B and let A be a closed subspace of X. The fibrewise topological space $X/_B A$ obtained from X by fibrewise collapsing A, as in §5, comes equipped with a closed section, given by $B = A/_B A \to X/_B A$.

Consider the fibrewise push-out of the cotriad

$$X \xleftarrow{s} B \xrightarrow{t} Y,$$

where X and Y are fibrewise pointed topological spaces over B with sections s and t. This fibrewise pointed topological space is denoted by $X \vee_B Y$ and called the *fibrewise pointed coproduct*. Since the projections are necessarily surjective we may regard $X \vee_B Y$ as a fibrewise quotient of the (fibrewise) topological coproduct $X +_B Y$. Note that the projection $\pi: X +_B Y \to X \vee_B Y$ is closed whenever the sections s and t are closed, since if $E \subset X$ is closed then the saturation $\pi^{-1}\pi E = E + ts^{-1}E$ is closed and similarly if $F \subset Y$ is closed. Moreover the fibres are finite, in this case, and so the projection π is proper.

The sections s and t also define a triad

$$X \xrightarrow{u} X \times_B Y \xleftarrow{v} Y,$$

where the components of u are (id_X, tp) and the components of v are (sq, id_Y). Since u, v are embeddings so is the fibrewise push-out

$$X \vee_B Y \to X \times_B Y;$$

the fibrewise collapse of the fibrewise push-out is denoted by $X \wedge_B Y$ and called the *fibrewise smash product*. Note that $X \wedge_B Y$ is fibrewise compact when X and Y are fibrewise compact, and fibrewise regular when X and Y are fibrewise regular, by (5.6).

Fibrewise pointed topological spaces where the section is closed play a special role, as in

Proposition (6.1). *Let X, Y and Z be fibrewise pointed topological spaces over B with closed sections. Then there exists a natural equivalence*

$$(X \wedge_B Z) \vee_B (Y \wedge_B Z) \to (X \vee_B Y) \wedge_B Z,$$

of fibrewise pointed topological spaces.

For consider the diagram shown below, where the functions are the obvious ones.

$$
\begin{array}{ccc}
X \times_B Z +_B Y \times_B Z & \longrightarrow X \wedge_B Z +_B Y \wedge_B Z \longrightarrow & (X \wedge_B Z) \vee_B (Y \wedge_B Z) \\
\downarrow & & \downarrow \\
(X +_B Y) \times_B Z & \longrightarrow (X \vee_B Y) \times_B Z \longrightarrow & (X \vee_B Y) \wedge_B Z
\end{array}
$$

The horizontals on the right are fibrewise quotients by definition. The upper horizontal on the left is a fibrewise coproduct of fibrewise quotients, therefore a fibrewise quotient. The lower horizontal on the left is the fibrewise product of the proper fibrewise function π with the identity, hence proper and so also a fibrewise quotient. The left-hand vertical is a fibrewise topological equivalence by the distributive law for the fibrewise topological product. The right-hand vertical is a bijection and hence, by commutativity of the diagram, a fibrewise topological equivalence.

Proposition (6.2). *Let X, Y and Z be fibrewise pointed topological spaces over B with closed sections. Suppose that either* (i) *X and Z are fibrewise locally compact and fibrewise regular, or* (ii) *X and Y (or Y and Z) are fibrewise compact and fibrewise regular. Then there exists a natural equivalence $(X \wedge_B Y) \wedge_B Z \to X \wedge_B (Y \wedge_B Z)$ of fibrewise pointed topological spaces.*

To prove this, under either hypothesis, consider the commutative diagram shown below, where the horizontals are the obvious functions and the right-hand vertical is induced by the identity at the left.

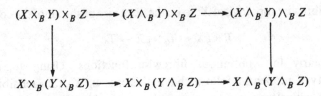

Under hypothesis (i) the upper horizontal on the left is a fibrewise quotient map, by (5.3), while that on the right is a fibrewise quotient map by definition. Thus the composition is a fibrewise quotient map. Similarly the lower horizontal composition is a fibrewise quotient map. It follows at once that the bijection on the right is a fibrewise topological equivalence. Under hypothesis (ii), with X and Y fibrewise compact and fibrewise regular, the projection $X \times_B Y \to X \wedge_B Y$ is proper, by (5.4), and so the upper horizontal on the left is fibrewise proper and so a fibrewise quotient map. Hence the upper horizontal composition is a fibrewise quotient map, moreover so is the lower horizontal composition as before. The result now follows at once.

We refer to the above two results as the distributive law and the associative law for the fibrewise smash product. The proof of our next result is similar to that of (6.2) and will therefore be left to serve as an exercise.

Proposition (6.3). *Let X and Y be fibrewise pointed topological spaces over B with closed sections, and let A be a closed subspace of X. If Y is fibrewise locally compact and fibrewise regular, or if A is fibrewise compact, then there exists a natural equivalence $(X \wedge_B Y)/_B(A \wedge_B Y) \to (X/_B A) \wedge_B Y$ of fibrewise pointed topological spaces.*

Suppose that we have an endofunctor Φ_B of the category of fibrewise topological spaces over B. For each fibrewise pointed topological space X over B the section $B \to X$ has a left inverse, i.e. the projection, hence the transform $\Phi_B(B) \to \Phi_B(X)$ has a left inverse. We define $\Phi_B^B(X)$ to be the fibrewise push-out of the cotriad

$$\Phi_B(X) \leftarrow \Phi_B(B) \to B,$$

and similarly for continuous fibrewise pointed functions. The endofunctor Φ_B^B of the category of fibrewise pointed topological spaces over B thus defined is called the *reduction* of the original endofunctor Φ_B.

For example, let T be a fibrewise locally compact and fibrewise regular space over B, and let T_0 be a closed subspace of T. As before define $\Phi_B(X)$, for each fibrewise topological X, to be the fibrewise push-out of the cotriad

$$T \times_B X \leftarrow T_0 \times_B X \rightarrow T_0$$

and similarly for continuous fibrewise functions. Using (6.3) and transitivity of quotients it follows that in this case $\Phi_B^B(X)$, for each fibrewise pointed X with closed section, is equivalent to the fibrewise smash product $(T/_B T_0) \wedge_B X$, and similarly for continuous fibrewise pointed functions. In particular the reduced fibrewise cone functor Γ_B^B is equivalent to $(B \times I) \wedge_B$ and the reduced fibrewise suspension functor Σ_B^B to $(B \times S) \wedge_B$ where I has basepoint 0 and $S = I/\dot{I}$ has basepoint $\dot{I}/\dot{I}$.

7. Relation with equivariant topology

The reader is assumed to be familiar with the basic theory of topological transformation groups as contained in [16] or [51], for example. We consider G-spaces, where G is a topological group. We regard a G-space X as fibrewise topological over its orbit space X/G. Then X is fibrewise open, in all cases, and is fibrewise compact whenever G is compact. Furthermore X is locally sliceable when G is a compact Lie group and X is completely regular.

Not a great deal can be said about G-spaces in general. We shall confine our discussion very largely to G-spaces which satisfy the following condition.

Definition (7.1). *The G-space X is* proper *if the continuous surjection* $\theta: G \times X \rightarrow X \times_{X/G} X$, *given by*

$$\theta(g, x) = (x, g \cdot x) \quad (g \in G, x \in X)$$

is proper.

Note that θ, in (7.1), is bijective if and only if the action is free. Thus θ is an equivalence if and only if X is a free proper G-space.

Note that for any G-space X the fibrewise graph

$$\Gamma: G \times X \rightarrow G \times X \times_{X/G} X$$

of the action is an embedding. If X is fibrewise Hausdorff over X/G the fibrewise graph is closed, by (2.10), and so Γ is proper. If in addition G is compact then the projection

$$\pi: G \times X \times_{X/G} X \rightarrow X \times_{X/G} X$$

is proper, hence $\theta = \pi\Gamma$ is proper. Thus we have

Proposition (7.2). *Let X be a G-space, where G is compact. If X is fibrewise Hausdorff then X is a proper G-space.*

In the other direction we have

Proposition (7.3). *Let X be a proper G-space, where G is Hausdorff. Then X is fibrewise Hausdorff.*

For the diagonal in the fibrewise product can be expressed as the composition

$$X \overset{\Gamma}{\to} G \times X \overset{\theta}{\to} X \times_{X/G} X,$$

where now Γ is the graph of the constant function, with a switch of factors. But Γ is a closed embedding, by (2.10), since G is Hausdorff. Therefore $\theta\Gamma$ is proper and so closed. Hence and from (2.7) we obtain (7.3). We conclude, therefore, that for compact Hausdorff G the G-space X is proper if and only if X is fibrewise Hausdorff.

Not every G-space is fibrewise regular over X/G. For example, take G to be the group of integers, with discrete topology, and take X to be the same set, with cofinite topology. Then X is a G-space, using addition of integers, but X is not regular, as a topological space, and so is not fibrewise regular, as a fibrewise topological space over the point-space X/G. However we can prove

Proposition (7.4). *Let X be a proper G-space. Then X is fibrewise regular over X/G.*

For let $x \in X_b$, where $b \in B = X/G$, and let U be a neighbourhood of x in X. The stabilizer G_x of x is compact, since

$$\theta^{-1}(x, x) = G_x \times \{x\},$$

and so there exists a neighbourhood N of G_x in G, and a neighbourhood V' of x in X such that $N \times V' \subset \theta^{-1}(U \times_B U)$. Now $N \times V'$ is a neighbourhood of $\theta^{-1}(x, x)$ and so there exists, since θ is closed, a neighbourhood V of x such that $\theta^{-1}(V \times_B V)$ is contained in $N \times V'$. Write $W = pV$, where $p: X \to B$ is the natural projection; then $X_W = GV$. I assert that $X_W \cap \text{Cl } V$ is contained in U. For suppose that $gy \in \text{Cl } V$, where $g \in G$ and $y \in V$. Then gV, being a neighbourhood of gy, meets V in z, say. Now $\theta(g, g^{-1}z) = (g^{-1}z, z) \in V \times_B V$, hence $g \in N$. But $y \in V$ and so $gy \in NV \subset U$, as required. This proves (7.4).

Proposition (7.5). *Let H be a subgroup of the topological group G. The translation action of G on G/H is proper if and only if H is compact.*

Here the orbit space of the action is trivial, since the action is transitive, and

$$\theta: G \times G/H \rightarrow G/H \times G/H.$$

If θ is proper then $\theta^{-1}(e, e) = H \times \{e\}$ is compact, where $e = [H] \in G/H$, and so H is compact. Conversely suppose that H is compact, and consider the diagram shown below.

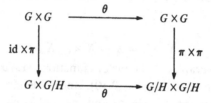

The upper θ is proper, since obviously G acts properly on itself by translation. Also the projection π is proper, since H is compact, and so $\pi \times \pi$ is proper. Thus $\theta(\mathrm{id} \times \pi) = (\pi \times \pi)\theta$ is proper and so the lower θ is proper, by (3.7), as required.

Proposition (7.6). *Let X be a free G-space. Then the division function $d: X \times_{X/G} X \rightarrow G$, given by*

$$\eta = d(\xi, \eta) \cdot \xi \quad (\xi, \eta \in X_b, \, b \in X/G),$$

is continuous if and only if the action of G is proper.

In this case θ is a continuous bijection, since the action is free. If θ is proper then θ is a homeomorphism and so d, being the first projection of the inverse of θ, is continuous. For the converse the same argument is used in reverse.

Proposition (7.7). *Suppose that X is a properly discontinuous G-space, i.e. a free proper G-space where G is discrete. Then X is fibrewise discrete over X/G.*

For the diagonal in $X \times_{X/G} X$ coincides with the inverse image $d^{-1}(e)$, where d is as in (7.6). Since $\{e\}$ is open in G the diagonal is open in $X \times_{X/G} X$ and so (7.7) follows from (1.23).

We now turn our attention to the situation where X is fibrewise topological over B and the topological group G acts fibrewise on X, so that a continuous function $\lambda: X/G \rightarrow B$ is induced by the projection.

We suppose that X is fibrewise non-empty, so that λ is surjective, and that X is fibrewise open, so that λ is open. Thus λ is a continuous open surjection. If we further assume that λ is injective then λ is a homeomorphism. When all these conditions are satisfied we describe X as a *special fibrewise G-space* over B, and prove

Proposition (7.8). *Let $\phi: X \to Y$ be a fibrewise G-map, where X and Y are special fibrewise G-spaces over B. If the action of G on Y is free and proper then ϕ is a G-equivalence.*

By (1.12) it is sufficient to show that the product

$$\mathrm{id} \times \phi: X \times_B X \to X \times_B Y$$

is a homeomorphism. By (7.6) the division function $d: Y \times_B Y \to G$ is continuous, since the action of G on Y is free and proper. Now an inverse of $\mathrm{id} \times \phi$ is given by the composition

$$X \times_B Y \underset{\Delta \times \mathrm{id}}{\to} X \times_B X \times_B Y \underset{\mathrm{id} \times \phi \times \mathrm{id}}{\to} X \times_B Y \times_B Y \underset{\mathrm{id} \times d}{\to} X \times G \underset{\theta}{\to} X \times_B X,$$

and so the result follows.

Corollary (7.9). *Let X be a special fibrewise G-space over B. Suppose that the action of G on X is free and proper. If X is sectionable (resp. locally sectionable) then X is trivial (resp. locally trivial) as a fibrewise G-space.*

For if $s: B \to X$ is a section then a fibrewise G-map $\phi: B \times G \to X$ is given by

$$\phi(b, g) = g \cdot s(b) \quad (b \in B, g \in G).$$

The local case is similar.

The *mixed product* of a right G-space X and a left G-space Y is the quotient space $X \times_G Y$ of $X \times Y$ with respect to the equivalence relation

$$(x, gy) \sim (xg, y) \quad (x \in X, y \in Y, g \in G).$$

Proposition (7.10). *Let G and H be topological groups. Let X be a right G-space, let Y be a left G-space and a right H-space, and let Z be a left H-space. Suppose that the action of G on the left of Y commutes with the action of H on the right of Y. Then G acts on the left of $Y \times_H Z$, and H acts on the right of $X \times_G Y$, so that the topological spaces $(X \times_G Y) \times_H Z$ and $X \times_G (Y \times_H Z)$ are equivalent.*

This result justifies the omission of brackets in the situation in question. The proof is straightforward and is therefore left as an exercise, as is

Proposition (7.11). *Let G and H be topological groups. Let X be a right G-space. Let Y be a left G-space and a right H-space, with the actions commuting. Then there is an induced action of H on the right of $X \times_G Y$ such that the topological spaces $(X \times_G Y)/H$ and $X \times_G (Y/H)$ are equivalent.*

This shows, for example, that if X is a right G-space and H is a subgroup of G then $X \times_G (G/H)$ is equivalent to $(X \times_G G)/H$ and hence to X/H, a result which is frequently useful.

By taking the mixed product with a given right G-space P a functor $P_\#$ is defined from the category of (left) G-spaces to the category of fibrewise topological spaces over P/G. Specifically, $P_\#$ transforms the G-space A into the mixed product $P \times_G A$, regarded as a fibrewise topological space over P/G through π_1/G, and similarly with morphisms. When P is a principal G-bundle over P/G (see §25) then $P_\#$ is just the associated bundle with fibre A. Note that if H acts on the right of A, and the action commutes with that of G, then $P_\#(A/H)$ and $(P_\# A)/H$ are equivalent, as fibrewise topological spaces over P/G, by (7.11).

To give an illustration, recall that for any topological space A the configuration spaces $A^{(n)}$ of A are defined as follows, for $n = 1, 2, \ldots$. Form the topological nth power A^n and consider the subspace A^n_+ consisting of n-tuples $(a_1, \ldots, a_n)$, where the a_i are all distinct. The symmetric group $S(n)$ acts on A^n by permuting factors. The action leaves A^n_+ invariant, and so the orbit space $A^n_+/S(n) = A^{(n)}$ is defined. If A is a G-space then so is A^n, moreover the actions of G and $S(n)$ commute. Since A^n_+ is invariant with respect to both actions we obtain that the configuration space $A^{(n)}$ is also a G-space.

Obviously there is a fibrewise version of this construction. Thus let X be fibrewise topological over B. Then the fibrewise configuration space $X_B^{(n)}$ is defined by forming the fibrewise topological nth power X_B^n, taking the subspace consisting of distinct n-tuples in each fibre, and passing to the fibrewise quotient. In particular if $X = P_\# A$, for some G-space A, then (7.11) shows that $X_B^{(n)} = P_\# A^{(n)}$. These fibrewise configuration spaces have been studied by Harvey [27]. For compact G there are some useful relationships between the topology of A and the fibrewise topology of $P_\# A$, as in the following three results.

Proposition (7.12). *Let P be a right G-space and A a left G-space, where G is compact. If A is compact then $P_\# A$ is fibrewise compact over $B = P/G$.*

For P is fibrewise compact, since G is compact. Also $B \times A$ is fibrewise compact, since A is fibrewise compact, and so $P \times A = P \times_B (B \times A)$ is fibrewise compact. Now the conclusion follows from (3.7).

Proposition (7.13). *Let P be a right G-space and A a left G-space, where G is compact Hausdorff. If A is Hausdorff then $P_\# A$ is fibrewise Hausdorff over $B = P/G$.*

For P is fibrewise Hausdorff, by (7.3), since G is compact Hausdorff. Also $B \times A$ is fibrewise Hausdorff, since A is Hausdorff, and so $P \times A = P \times_B (B \times A)$ is fibrewise Hausdorff. Now the conclusion follows from (3.21).

Proposition (7.14). *Let P be a right G-space and A a left G-space, where G is compact Hausdorff. If A is locally compact and Hausdorff then $P_\# A$ is fibrewise locally compact and fibrewise regular over B.*

For P is fibrewise compact, since G is compact. Also $B \times A$ is fibrewise locally compact, since A is locally compact, and so $P \times A = P \times_B (B \times A)$ is fibrewise locally compact. Also $P \times A$ is fibrewise Hausdorff, as in the proof of (7.11), and so fibrewise regular, by (3.22). Now the conclusion follows using (3.23).

We describe an equivalence relation R on the G-space X as *equivariant* if $\xi R\eta$ implies $g\xi Rg\eta$ for all $g \in G$. When this condition is satisfied the set X/R of equivalence classes can be given the structure of a G-set so that the projection $\pi: X \to X/R$ is equivariant. Then we give X/R the quotient topology determined by π, so that X/R is the quotient space. In general the action of G on X/R is not continuous. By the results proved in §5 above, however, the action will be continuous if either (i) π is a proper map or (ii) G is locally compact.

For an example of (i), let X be a G-space and let A be a closed invariant subspace of X. By (5.4), if A is compact then the projection $\pi: X \to X/A$ is proper, and so X/A is a G-space with π equivariant. If G is compact, moreover, then the projection of X/A onto $(X/A)/G$ is proper, and hence $(X/A)/G$ is equivalent to $(X/G)/(A/G)$.

For an example of (ii), let T be a G-space where G is locally compact, and let T_0 be a closed invariant subspace of T. Then an endofunctor Φ of the category of G-spaces is defined which transforms each G-space X

into the equivariant push-out $\Phi(X)$ of the cotriad

$$T \times X \leftarrow T_0 \times X \rightarrow T_0,$$

and similarly for morphisms. Moreover the orbit space $(\Phi(X))/G$ can be identified with the ordinary push-out of the cotriad

$$T \times_G X \leftarrow T_0 \times_G X \rightarrow T_0/G.$$

In particular, suppose that G acts trivially on T. Then $(\Phi(X))/G$ can be identified with $\Phi(X/G)$, where G acts trivially on X/G. For example, $(\Gamma(X))/G$ is equivalent to $\Gamma(X/G)$ and $(\Sigma(X))/G$ is equivalent to $\Sigma(X/G)$.

Let T be a G-space, with G compact, and let T_0 be a closed invariant subspace. Let X be a pointed G-space with closed basepoint x_0. Then the endofunctor Φ, as before, transforms X into a G-space $\Phi(X)$, containing $\Phi(x_0) = T$ as a closed invariant subspace. The quotient space $\Phi(X)/\Phi(x_0)$ is then a pointed G-space which we denote by $\Phi^*(X)$. In this way an endofunctor of the category of pointed G-spaces is defined, the reduced endofunctor of Φ. The reduced cone Γ^* and the reduced suspension Σ^* are special cases.

To link up the results given here with those given previously, let P be a right G-space and write $P/G = B$. Then P determines a functor $P_\#$ from the category of (pointed) left G-spaces to the category of fibrewise (pointed) topological spaces over B, by taking the mixed product. It is easy to see that $P_\#$ respects coproducts and products, also that $P_\#$ transforms proper morphisms into proper fibrewise functions, and in particular transforms closed invariant subspaces into closed subspaces.

For example, consider the sphere

$$S^n = \{x \in \mathbb{R}^{n+1} : \|x\| = 1\},$$

with basepoint the point $p_0 = (1, 0, \ldots, 0)$. The orthogonal group $O(n+1)$ acts transitively on S^n, in the usual way. We first regard S^n as a pointed $O'(n)$-space, where $O'(n)$ is the stabilizer of p_0. I assert that $\Gamma^*(S^n)$ is equivalent to the $(n+1)$-ball

$$B^{n+1} = \{x \in \mathbb{R}^{n+1} : \|x\| \leqslant 1\},$$

as a pointed $O'(n)$-space. In fact an equivalence $\theta : \Gamma^*(S^n) \rightarrow B^{n+1}$ is given by

$$\theta(t, z) = (tz + (1-t)p_0),$$

where $t \in I$ and $z \in S^n$. Next we regard S^n as a pointed $O'(n-1)$-space, where $O'(n-1)$ is the stabilizer of the pole $(0, \ldots, 0, 1)$ as well as p_0.

I assert that $\Sigma^*(S^n)$ is equivalent to the $(n+1)$-sphere

$$S^{n+1} = \{x \in \mathbb{R}^{n+2} : \|x\| = 1\},$$

as an $O'(n-1)$-space. In fact an equivalence $\phi : \Sigma^*(S^n) \to S^{n+1}$ is given by

$$\phi(t, z) = \begin{cases} p_+(2tz + (1-2t)p_0) & (t \leqslant 1/2) \\ p_-((2-2t)z + (2t-1)p_0) & (t \geqslant 1/2) \end{cases}$$

where p_+, p_- denote the orthogonal projections of the equatorial n-space onto the upper and lower hemisphere respectively.

We conclude from this that if P is a right $O'(n)$-space, as above, then the *reduced* fibrewise cone of $P_{\#} S^n$ is equivalent to $P_{\#} B^{n+1}$, as a fibrewise pointed topological space, while if P is a right $O'(n-1)$-space then the *reduced* fibrewise suspension of $P_{\#} S^n$ is equivalent to $P_{\#} S^{n+1}$ in the same sense. These conclusions have some relevance to the questions considered at the end of §26.

Exercises

1. Let X be a fibrewise topological space over the cylinder $I \times A$, where A is any topological space. Suppose that X is trivial over $[0, \frac{1}{2}] \times A$ and over $[\frac{1}{2}, 1] \times A$. Show that X is trivial over $I \times A$.

2. Let X be a fibrewise topological space over the closed unit interval I. If X is locally trivial show that X is trivial. Generalize to the case of the closed n-cube I^n.

3. Let $\phi : X \to Y$ be an open fibrewise surjection, where X and Y are fibrewise topological over B. Suppose that the inverse image with respect to $\phi \times \phi$ of the diagonal of Y is closed in $X \times_B X$. Show that Y is fibrewise Hausdorff.

4. Let $B = \{0, 1\}$, with the topology given by supersets of 0; let $X = \{0, 1, 2\}$, with similar topology; and let the projection p be given by $p(0) = 0$ and $p(1) = p(2) = 1$. Show that X is fibrewise discrete but not fibrewise Hausdorff.

5. Denote the set of integers with discrete topology by $\mathbb{Z}$, with cofinite topology by $\check{\mathbb{Z}}$. Show that $\check{\mathbb{Z}} \times \mathbb{Z}$ is not fibrewise normal over $\check{\mathbb{Z}}$, under the first projection, although $\mathbb{Z}$ is normal. (Hint: consider the diagonal Δ and superdiagonal Δ', where $\Delta = \{t, t\}$, $\Delta' = \{t, t+1\}$ $(t = 0, \pm 1, \ldots).$)

6. Let $\phi : X \to Y$ be a fibrewise function, where X and Y are fibrewise topological over B. Suppose that X is fibrewise Hausdorff and that the fibrewise graph of ϕ in $X \times_B Y$ is fibrewise compact. Show that ϕ is continuous.

7. Let X be a fibrewise compact subspace of $B \times \mathbb{R}^n$ $(n \geqslant 0)$, where B is paracompact. Show that X is fibrewise bounded in the sense that there exists a continuous function $\alpha : B \to \mathbb{R}$ such that X_b is bounded by $\alpha(b)$ for each point b of B.

8. Let $\phi: X \to Y$ be a continuous fibrewise function, where X and Y are fibrewise discrete over B. Show that ϕ is a local homeomorphism.

9. Let X be fibrewise topological over B and let A be an open subspace of X. If A is fibrewise open show that the natural projection $\pi: X \to X/_B A$ is open.

10. Take $B = \{0, 1\}$, with $\{0\}$ open and $\{1\}$ not, and take $X = \{0, 1, 2\}$ with $X_0 = \{0, 2\}$ and $X_1 = \{1\}$. Show that although X, with the fibrewise indiscrete topology, is fibrewise compact there exists no closed fibrewise embedding of X in $B \times T$ for any topological space T.

11. Let X be fibrewise compact over B. Let $\mathscr{F}$ be a b-filter on X, where $b \in B$. Suppose that the adherence set of $\mathscr{F}$ consists of precisely one point. Show that $\mathscr{F}$ is convergent.

12. Let $\{X_r\}$ be a family of fibrewise compact spaces over B. Show that the fibrewise topological coproduct $\coprod_B X_r$ is fibrewise compact if and only if the family $\{p_r X_r\}$ of subsets of B is locally finite.

13. Let $\phi: X \to Y$ be a proper fibrewise function, where X and Y are fibrewise topological over B. Show that if X is fibrewise Hausdorff and fibrewise regular then so is ϕX.

14. Let X be fibrewise topological over B. Let R be the fibrewise equivalence relation on X given by $\xi R \eta$, where $\xi, \eta \in X_b$, $b \in B$, if there exists a neighbourhood W of b such that $X_W \cap \bar{\xi} = X_W \cap \bar{\eta}$. Show that the projection $\pi: X \to X/R$ is open and closed. Also show that X/R is fibrewise T_0.

15. Let X be a G-space, where G is a topological group. Show that X is fibrewise discrete over X/G if and only if the action is discontinuous, in the sense that each point of X admits a neighbourhood V such that $\xi \neq \eta$, where $\xi, \eta \in V$, implies $g\xi \neq \eta$ for all $g \in G$.

16. Let X be a G-space, where G is discrete, and suppose that X is fibrewise Hausdorff over X/G. Show that the action is proper if and only if for each pair of points ξ, η in the same orbit of X there exist neighbourhoods U of ξ, V of η such that the subset

$$\{g \in G: gU \cap V \neq \varnothing\}$$

is finite.

17. Let X and Y be G-spaces, where G is discrete. Suppose that X is fibrewise Hausdorff over X/G and that Y is fibrewise Hausdorff over Y/G. If G acts properly on X show that G acts properly on $X \times Y$.

II

Further fibrewise topology

8. Fibrewise compactification

Of the many methods of compactification which are treated in the literature, two of the most important are the Wallman, of which the Stone–Čech may be regarded as a modification, and the Alexandroff (or one-point) compactification. We begin this second chapter by showing how to construct fibrewise versions of these two methods, which are very different in character, taking the Wallman method first.

As before we work over a topological base space B. Let X be fibrewise topological over B. We describe a subset M of X as *b-closed*, where $b \in B$, if $X_W \cap M$ is closed in X_W for some neighbourhood W of b. Thus it is sufficient, but not necessary, for M to be closed in X. For example X_W is b-closed for each neighbourhood W of b. Note that finite unions and finite intersections of b-closed sets are again b-closed.

By a *b-closed filter* on X, where $b \in B$, we mean a filter $\mathscr{F}$ on X consisting entirely of b-closed sets, which is tied to b in the sense that $X_W \in \mathscr{F}$ for each neighbourhood W of b. This is in line with the terminology used in §4 except that there the filters consist of unrestricted subsets of X. By a *refinement* of a b-closed filter $\mathscr{F}$ we mean a b-closed filter $\mathscr{F}'$ such that each member of $\mathscr{F}$ is also a member of $\mathscr{F}'$. By a *b-closed ultrafilter* we mean a b-closed filter which does not admit any strict refinement. When X is fibrewise R_0-space as defined in (2.2), the filter consisting of all b-closed sets containing a given point $x \in X_b$ is a b-closed ultrafilter, as can easily be seen.

Proposition (8.1). *Let $\mathscr{F}$ be a b-closed ultrafilter on X. If $M \cup N \in \mathscr{F}$ where M and N are b-closed, then $M \in \mathscr{F}$ or $N \in \mathscr{F}$.*

For suppose that $M \cup N \in \mathscr{F}$ but $M \notin \mathscr{F}$. Let $\mathscr{F}'$ be the family of b-closed sets N' of X such that $M \cup N' \in \mathscr{F}$. Then $\mathscr{F}'$ is a b-closed filter which refines $\mathscr{F}$. Therefore $N \in \mathscr{F}$ since otherwise the refinement would be strict.

In what is to follow we denote by $[M]$ the set of b-closed ultrafilters which contain a given b-closed set M as a member. We have the relations

$$[M_1 \cup M_2] = [M_1] \cup [M_2], \qquad [M_1 \cap M_2] = [M_1] \cap [M_2],$$

where M_1 and M_2 are b-closed.

Now consider the fibrewise set X' over B of which the fibre X'_b over a given point b of B is the set of b-closed ultrafilters on X. The *fibrewise Wallman compactification* of X is defined to be the fibrewise set X' with the following fibrewise topology: when H is a subset of X' and $b \in B$ then $\xi \in X'_b$ is an adherence point of H if $\xi \in [M]$ for each b-closed set M of X for which $[M]$ contains H.

From now on we assume that X is a fibrewise R_0-space. Then a fibrewise injection $\rho \colon X \to X'$ is defined which assigns to each point $x \in X_b$, where $b \in B$, the b-closed ultrafilter consisting of all b-closed sets which contain x. Clearly $\rho^{-1}[M] = M$, for each b-closed set M of X. It follows that ρ is an embedding. Moreover if M is a b-closed set of X then $\xi \in X'_b$ is an adherence point of ρM if and only if $\xi \in [M]$. In particular, taking $M = X$, we see that ρX is dense in X'.

We shall now show, in two steps, that X' is fibrewise compact, and thus is entitled to be described as a fibrewise compactification of X. The appropriate notation for the fibrewise Wallman compactification would be $\omega_B X$, but we shall retain X' for simplicity.

First suppose that we have a b-closed filter $\mathscr{F}$ on X. I assert that $\mathscr{F}$ admits an adherence point $\xi \in X'_b$. This is obvious if the common intersection C of all the members of $\mathscr{F}$ meets X_b, since in that case we can take ξ to be any point of $X_b \cap C$. If, however, C does not meet X_b then we can use Zorn's lemma to obtain a b-closed ultrafilter $\mathscr{G}$ which refines $\mathscr{F}$. If $H \in \mathscr{F}$ then $\mathscr{G} \in [H]$, since $H \in \mathscr{G}$, so that $\mathscr{G}$ adheres to $\mathscr{F}$. Thus X' contains an adherence point of $\mathscr{F}$ in either case.

Now suppose instead that we have a b-filter $\mathscr{F}'$ on X'. Consider the family $\mathscr{F}$ of b-closed sets M of X such that $[M]$ contains some member H of $\mathscr{F}'$. Now $M_1 \cap M_2 \in \mathscr{F}$ whenever $M_1 \in \mathscr{F}$ and $M_2 \in \mathscr{F}$, since then $H_1 \subset [M_1]$ and $H_2 \subset [M_2]$, where $H_1, H_2 \in \mathscr{F}$, and so $H_1 \cap H_2 \subset [M_1] \cap [M_2] = [M_1 \cap M_2]$. Thus $\mathscr{F}$ is a b-closed filter on X and so admits an adherence point $\xi \in X'_b$, by the first step in the argument. Suppose then that $H \subset [M]$, where $H \in \mathscr{F}'$ and M is b-closed. Then ξ is an adherence point of M, hence $\xi \in [M]$ and hence ξ is an adherence point of H. We conclude, therefore, that ξ is an adherence point of $\mathscr{F}'$, and so X' is fibrewise compact.

In fact, fibrewise compactness in itself is not particularly useful unless it is accompanied by the fibrewise Hausdorff property. First observe that if

X' is fibrewise Hausdorff then X is fibrewise Hausdorff, by (2.8), also X' is fibrewise normal, by (3.27). We prove

Proposition (8.2). *Suppose that X is fibrewise normal (as well as fibrewise R_0). Then the fibrewise Wallman compactification X' of X is fibrewise Hausdorff.*

For let $\xi, \eta \in X'_b$ be distinct, where $b \in B$. If each of the intersections $X_W \cap M \cap N$ is non-empty, where W runs through the neighbourhoods of b and M, N run through the members of ξ, η, respectively, then the intersections generate a strict refinement of ξ or η, contrary to the ultrafilter condition. We can choose W, M and N, therefore, so that $X_W \cap M \cap N$ is empty. Since X is fibrewise normal there exists a neighbourhood $W' \subset W$ of b and open sets U, V of $X_{W'}$ such that $X_{W'} \cap M \subset U$, $X_{W'} \cap N \subset V$ and $X_{W'} \cap U \cap V$ is empty. Then $X_{W'} - X_{W'} \cap U$ and $X_{W'} - X_{W'} \cap V$ are b-closed sets of X, and so $[X_{W'} - X_{W'} \cap U]$ and $[X_{W'} - X_{W'} \cap V]$ are b-closed sets of X', of which the complements, in $X'_{W'}$, are neighbourhoods of ξ and η, respectively, which are disjoint.

Most, if not all, of the more important types of compactification can be regarded as modifications of the Wallman compactification, and fibrewise versions of these can be constructed in a similar fashion to the above. For example, the fibrewise Stone–Čech compactification, for fibrewise completely regular spaces, can be so constructed, using filters generated by subsets which are functionally b-closed, in an obvious sense. In fact there are various results in the literature which may be regarded as fibrewise versions of the Stone–Čech compactification, under a variety of conditions which unfortunately are unnecessarily restrictive from our point of view. For example Whyburn [63] requires complete regularity for the topological space in question; see also the more recent papers of Cain [11], Dyckhoff [21] and others.

If such restrictions are acceptable then in fact, a rather simple way of deriving a fibrewise compactification from a compactification is as follows. Let X be fibrewise topological over B. Suppose that X, as an ordinary space, satisfies the conditions under which a particular type of compactification $\hat{X}$ of X is defined. Thus $\hat{X}$ is compact and X is embedded in $\hat{X}$ as a dense subspace. A fibrewise embedding of X in $B \times \hat{X}$ is given by the projection in the first factor, by the inclusion in the second. Since $B \times \hat{X}$ is fibrewise compact so is every closed subspace. We can therefore take the closure, in $B \times \hat{X}$, of the image of X as the fibrewise compactification of this particular type.

An interesting question to which the answer at present appears to be unknown is the following. Let X be fibrewise topological over B. If there exists a closed fibrewise embedding of X in $B \times T$ for some compact T then X is fibrewise compact. Conversely, suppose that X is fibrewise compact: does there exist a closed fibrewise embedding of X in $B \times T$ for some compact T? By taking T to be the Wallman compactification of X (as an ordinary space) we see that the answer is affirmative when X is normal, at least. The question has been considered by Dyckhoff [21].

We now turn from the Wallman to the Alexandroff method of compactification. Let X be fibrewise topological over B, as before. Consider the fibrewise set $X +_B B = X_B^+$ with section the insertion $s: B \rightarrow X_B^+$ and with the fibrewise topology generated by the fibrewise neighbourhood system in which the neighbourhood b-filter $(b \in B)$ of a point of the fibre over b is generated by (i) for points of X_b, the neighbourhood b-filter in X and (ii) for the point $s(b)$, sets of the form $(X_W - K) + sW$, where W is a neighbourhood of b and where K is a closed set of X_W which is fibrewise compact over W. In particular, taking $W = B$ and $K = \varnothing$, we see that X is open in X_B^+, and that s is a closed section.

I assert that the fibrewise topological space X_B^+ over B thus defined is fibrewise compact. To see this, let b be a point of B and let $\mathscr{F}$ be a b-ultrafilter on X_B^+. Suppose, to obtain a contradiction, that $\mathscr{F}$ has no limit point. Then $s(b)$ is not a limit point and so there exists a neighbourhood W of b in B and a closed set K of X_W such that $X_W - K + sW \notin \mathscr{F}$. Since $X_W + sW \in \mathscr{F}$ it follows that $K \in \mathscr{F}$, by (4.1). Also each point x of X_b is not a limit point either, and so there exists a neighbourhood U_x of x in X_W such that $U_x \notin \mathscr{F}$. The family $\{U_x : x \in K_b\}$ constitutes a covering of the compact K_b by open sets of X_W. Since K is fibrewise compact over W there exists, by (3.3), a neighbourhood $W' \subset W$ of b such that a finite subfamily of the $\{U_x\}$ covers $X_{W'}$. Now $X_{W'} + sW' \in \mathscr{F}$, since $\mathscr{F}$ is tied to b, and so

$$K \cap X_{W'} = K \cap (X_{W'} + sW') \in \mathscr{F}.$$

Therefore $U_x \cap K \cap X_{W'} \in \mathscr{F}$, for some $x \in K_b$, by (4.1) again. But $U_x \notin \mathscr{F}$ and so we have our contradiction.

In view of this result we refer to X_B^+ as the *fibrewise Alexandroff compactification* of X. Note that if X is fibrewise Hausdorff over B then $K \subset X_W$ is closed in X_W whenever K is fibrewise compact over W, since X_W is fibrewise Hausdorff over W.

Now suppose that X is fibrewise locally compact and fibrewise Hausdorff over B. Then X_B^+ is fibrewise Hausdorff also. For since X is open in X_B^+ it is sufficient to show that each point x of X_b can be separated from $s(b)$, for each point b of B. Now since X is fibrewise locally compact there exists a neighbourhood W of b and a neighbourhood U of x in X_W such that $X_W \cap \bar{U}$ is fibrewise compact over W. Then U and $(X_W - X_W \cap \bar{U}) + sW$ are disjoint neighbourhoods of x and $s(b)$, respectively. Thus X_B^+ is fibrewise Hausdorff, as asserted.

Returning to the general case we prove

Proposition (8.3). *Let X be fibrewise Hausdorff over B and let A be an open subspace of X. Let $\phi: A \to Y$ be a proper fibrewise function, where Y is fibrewise topological over B. Then the fibrewise function $\psi: X \to Y_B^+$ is continuous, where ψ is fibrewise constant on $X - A$ and is given by ϕ on A.*

It is sufficient to show that the inverse image under ψ of a fibrewise subbasic neighbourhood of a point of the fibre over b is open for each point b of B. In the case of a point $y \in Y_b$, a fibrewise subbasic neighbourhood is just a neighbourhood V of y in Y, and $\psi^{-1}V = \phi^{-1}V$ which is open in A and hence in X. In the case of the point $t(b)$, a fibrewise subbasic neighbourhood is of the form $Y_W - L + tW$, where W is a neighbourhood of b and L is a closed set of Y_W which is fibrewise compact over W. Now $\phi^{-1}L$ is fibrewise compact over W, since ϕ is proper, and so closed in X_W, since X_W is fibrewise Hausdorff over W. Hence

$$\psi^{-1}(Y_W - L + tW) = X_W - \phi^{-1}L$$

is open in X_W and so open in X, as required. This completes the proof.

At the set-theoretic level, each fibrewise function $\phi: X \to Y$, where X and Y are fibrewise sets over B, determines a fibrewise pointed function $\phi^+: X_B^+ \to Y_B^+$, and vice versa. Suppose that $\phi: X \to Y$ is a proper fibrewise function, where X and Y are fibrewise topological spaces; then (8.3) shows that ϕ^+ is continuous provided X_B^+ is fibrewise Hausdorff. We deduce

Proposition (8.4). *Let X be fibrewise locally compact and fibrewise Hausdorff over B. Let A be a fibrewise compact subspace of X. Then there exists a natural equivalence*

$$(X/_A A)_B^+ \to X_B^+ /_B A_B^+$$

of fibrewise pointed topological spaces.

For A_B^+ is a closed subspace of X_B^+, since A is a closed subspace of X. Hence the natural projection $X_B^+ \to (X/_B A)_B^+$ induces a continuous fibrewise bijection

$$X_B^+ /_B A_B^+ \to (X/_B A)_B^+.$$

Now $X_B^+ /_B A_B^+$ is fibrewise compact, by (3.7), since X_B^+ is fibrewise compact. Also $X /_B A$ is fibrewise locally compact and fibrewise Hausdorff, by (3.23), (5.5) and (5.6), since X is fibrewise locally compact and fibrewise Hausdorff and since A is fibrewise compact. Therefore $(X /_B A)_B^+$ is fibrewise Hausdorff and so the continuous bijection is an equivalence, by (3.20).

Proposition (8.5). *Let X and Y be fibrewise locally compact and fibrewise Hausdorff over B. Then there exist natural equivalences*

$$X_B^+ \vee_B Y_B^+ \to (X +_B Y)_B^+, \qquad X_B^+ \wedge_B Y_B^+ \to (X \times_B Y)_B^+$$

of fibrewise pointed topological spaces.

To prove the first part of (8.5) consider the standard insertions

$$X \to X +_B Y \leftarrow Y.$$

Since these are both proper they induce continuous fibrewise pointed functions

$$X_B^+ \to (X +_B Y)_B^+ \leftarrow Y_B^+.$$

The fibrewise coproduct

$$X_B^+ \vee_B Y_B^+ \to (X +_B Y)_B^+$$

of these fibrewise pointed functions is a continuous bijection and hence an equivalence of fibrewise pointed topological spaces, by (3.20), since $X_B^+ \vee_B Y_B^+$ is fibrewise compact and $(X +_B Y)_B^+$ is fibrewise Hausdorff.

To prove the second part consider the fibrewise surjection

$$\xi: X_B^+ \times_B Y_B^+ \to (X \times_B Y)_B^+$$

which is given by the identity on $X \times_B Y$ and by the fibrewise constant function on $X_B^+ \vee_B Y_B^+$. Note that the domain of ξ is fibrewise compact and fibrewise Hausdorff, since X_B^+ and Y_B^+ are fibrewise compact and fibrewise Hausdorff, as is the codomain of ξ. The inverse images of the fibrewise compact subsets of $X \times_B Y$ are fibrewise compact and therefore closed in the domain. The inverse images of the open sets of $X \times_B Y$ are open in the domain since $X \times_B Y$ is open in $X_B^+ \times_B Y_B^+$. Hence ξ is continuous and so induces a continuous fibrewise bijection

$$\eta: X_B^+ \wedge_B Y_B^+ \to (X \times_B Y)_B^+.$$

Since the domain of η is fibrewise compact, by (4.5), and the codomain is fibrewise Hausdorff we conclude, using (4.19), that η is a fibrewise pointed topological equivalence.

Next let us consider fibrewise Alexandroff compactification from the equivariant point of view. Let X be fibrewise Hausdorff over B and let

G be a compact Hausdorff group acting fibrewise on X. Now $G \times X$ is open in $G \times X_B^+$, since X is open in X_B^+, also the action is proper, since G is compact. Hence by (8.3), we can extend the fibrewise action of G to X_B^+ so that points of the canonical section are left fixed. We prove

Proposition (8.6). *Let X be fibrewise locally compact and fibrewise Hausdorff over B. Let G be a compact Hausdorff group acting fibrewise on X. Then X_B^+/G is equivalent to $(X/G)_B^+$, as a fibrewise pointed topological space.*

For since G is compact the natural projection $\pi: X \to X/G$ is proper and so determines a continuous fibrewise pointed function

$$\pi^+: X_B^+ \to (X/G)_B^+ .$$

Also π^+ is invariant with respect to the action, since π is invariant, and so π^+ induces a continuous fibrewise function

$$\rho: X_B^+/G \to (X/G)_B^+ .$$

By inspection ρ is bijective. Now X_B^+ is fibrewise compact, by construction, and so X_B^+/G is fibrewise compact, by (3.7). Also X/G is fibrewise Hausdorff, by (3.21), and is fibrewise locally compact, by (3.15) and (3.23). Therefore $(X/G)_B^+$ is fibrewise Hausdorff and so ρ is an equivalence, by (3.20).

Recall from §7 that if P is a right G-space and A a left G-space, where G is a topological group, then the mixed product $P \times_G A = P_{\#} A$ is defined and regarded as a fibrewise topological space over $B = P/G$. Suppose that G is compact Hausdorff and that A is locally compact Hausdorff, so that the action of G on A extends to an action on the Alexandroff compactification A^+, leaving the additional point $*$ fixed. I assert that $P_{\#}(A^+)$ and $(P_{\#} A)_B^+$ are equivalent, as fibrewise pointed topological spaces over B.

To see this, note first that $P \times A$ is open in $P \times A^+$, since A is open in A^+. By (8.3), therefore, the inclusion of $P \times A$ in $P \times A^+$ admits a continuous fibrewise pointed extension

$$\xi: P \times A^+ \to (P \times A)_B^+ .$$

Now ξ is bijective, by inspection. Moreover ξ is equivariant and so induces a continuous fibrewise bijection

$$\xi/G: P \times_G A^+ \to (P \times A)_B^+/G .$$

Now the projection $\pi: P \times A \to P \times_G A$ is proper, since G is compact, and so induces a continuous fibrewise pointed function

$$\pi^+: (P \times A)_B^+ \to (P \times_G A)_B^+.$$

Moreover π^+ is equivariant, and so induces

$$\pi^+/G: (P \times A)_B^+/G \to (P \times_G A)_B^+.$$

Composing this with ξ/G and converting to the $P_\#$ notation we obtain a continuous fibrewise pointed bijection

$$P_\# A^+ \to (P_\# A)_B^+.$$

Now the domain here is fibrewise compact, by (7.12), since A^+ is compact. Also the domain is fibrewise Hausdorff, since $P_\# A$ is fibrewise locally compact and fibrewise Hausdorff, by (7.13) and (7.14). By (3.20), therefore, we have an equivalence, as asserted.

We conclude this section with some observations about fibrewise vector spaces; the connection with fibrewise Alexandroff compactification will soon become apparent. We work over a topological field **k**, either the real numbers, the complex numbers or the quaternions, and do not impose any dimensionality restrictions.

Let B be a topological space, as before. By a *fibrewise vector space* over B I mean a fibrewise pointed topological space X over B such that each fibre X_b ($b \in B$) is equipped with vector space structure, subject to the following three conditions. First $s(b)$ is required to be the zero vector of X_b, where $s: B \to X$ is the section. Secondly fibrewise vector addition $X \times_B X \to X$ is required to be continuous. Thirdly fibrewise scalar multiplication $L \times_B X \to X$ is required to be continuous, where $L = B \times \mathbf{k}$ is the product space. These conditions are satisfied, of course, in the case in which X is a vector bundle, but in our case there is no requirement of local triviality.

By a *fibrewise homomorphism* we mean a continuous fibrewise function $\phi: X \to Y$, where X and Y are fibrewise vector spaces over B, such that $\phi_b: X_b \to Y_b$ is a homomorphism for each point b of B. With this notion of morphism the fibrewise vector spaces form a category. The equivalences of the category are called *fibrewise isomorphisms*.

Note that the fibrewise topological product $X \times_B Y$ of fibrewise vector spaces over B is also a fibrewise vector space, with the direct sum vector space structure in each fibre.

Let X be a fibrewise vector space over B. Let X' be the complement of the zero section, so that each fibre X_b' consists of the non-zero vectors of X_b. Identify points of X_b' which differ by a non-zero scalar multiple. The fibrewise quotient space of X' thus obtained is called the *fibrewise projective space* associated with X and denoted by $P(X)$. Note that if A is a subspace of X, with A_b a subspace of X_b in the vector space sense for each $b \in B$, then $P(A)$ can be regarded as a subspace of $P(X)$, so that the fibrewise collapse $P(X)/_B P(A)$ is defined, as a fibrewise pointed topological space over B.

This brings us back to the fibrewise Alexandroff compactification. For consider the complement $(L \times_B X)'$ of the zero-section in the fibrewise topological product of $L = B \times \mathbf{k}$ and a fibrewise vector space X. A continuous fibrewise surjection

$$(L \times_B X)' \to X_B^+$$

is defined, by sending $((b, t), \xi)$ into $t^{-1}\xi$ for $t \neq 0$ and $\xi \in X_b$, into $s(b)$ for $t = 0$ and $\xi \in X_b'$. Moreover since the image is invariant under fibrewise scalar multiplication we have an induced continuous fibrewise surjection

$$\pi: P(L \times_B X) \to X_B^+.$$

Now π is fibrewise constant on $P(X)$, where X is embedded in $L \times_B X$ through

$$\xi \to ((b, 1), \xi) \quad (\xi \in X_b, b \in B),$$

and so we obtain a continuous fibrewise pointed surjection

$$\rho: P(L \times_B X)/_B P(X) \to X_B^+.$$

When $P(L \times_B X)$ is fibrewise compact and X_B^+ is fibrewise Hausdorff we can use (3.20) to conclude that ρ is a fibrewise topological equivalence.

For example, suppose that X is a fibrewise direct summand of $nL = B \times \mathbf{k}^n$ for some n, as is the case when B is compact Hausdorff and X is a finite-dimensional vector bundle over B. Then $L \times_B X$ is a fibrewise direct summand of $(n + 1)L = B \times \mathbf{k}^{n+1}$ and so $P(L \times_B X)$ is a fibrewise retract of $B \times P(\mathbf{k}^{n+1})$. The latter is fibrewise compact, since $P(\mathbf{k}^{n+1})$ is compact, and so $P(L \times_B X)$ is fibrewise compact by (3.7). Also X is fibrewise locally compact, by (3.13), and fibrewise Hausdorff, by (2.8), since $\mathbf{k}^n$ is locally compact and Hausdorff. Therefore X_B^+ is fibrewise Hausdorff and so is equivalent to $P(L \times_B X)/_B P(X)$, by (3.20).

9. The fibrewise mapping-space

We now turn to the problem of constructing an adjoint to the fibrewise topological product. This means, in effect, assigning a satisfactory fibrewise topology to the fibrewise set†

$$\text{map}_B(X, Y) = \coprod_{b \in B} \text{map}(X_b, Y_b),$$

where X and Y are fibrewise topological over B. In other words we need a fibrewise generalization of the compact-open topology. This can be defined as follows.‡ Let $(K, V; W)$, where $W \subset B$, $K \subset X_W$ and $V \subset Y_W$, denote the set of continuous functions $\psi \colon X_\beta \to Y_\beta$ $(\beta \in W)$ such that $\psi K_\beta \subset V_\beta$. We describe such a subset $(K, V; W)$ of $\text{map}_B(X, Y)$ as *fibrewise compact-open* if W is open in B, K is fibrewise compact over W, and V is open in Y_W. For each continuous function $\phi \colon X_b \to Y_b$, where $b \in B$, the fibrewise compact-open sets to which ϕ belongs form the fibrewise subbasis of a b-filter. Thus a fibrewise neighbourhood system is defined and we call the fibrewise topology generated thereby the *fibrewise compact-open* topology.

Proposition (9.1). *Let X and Y be fibrewise topological spaces over B. Then $\text{map}_B(X, B)$ is equivalent to B, and $\text{map}_B(B, Y)$ is equivalent to Y, as a fibrewise topological space.*

To prove the first statement we have to show that the projection $\text{map}_B(X, B) \to B$ is a homeomorphism. Since the projection is obviously a continuous bijection we only have to show that the projection is open and for this it is sufficient to show that the projection of a fibrewise compact-open set is open. Now if $(K, V; W)$ is fibrewise compact-open, where $V \subset W$ and $K \subset X_W$ is fibrewise compact over W, then the projection is $V \cup (W - pK)$ which is open since pK is closed in W.

To prove the second statement we consider the fibrewise bijection $\alpha \colon \text{map}_B(B, Y) \to Y$, given by evaluation. If $V \subset Y$ is open then $\alpha^{-1}(V) = (B, V; W)$, which is fibrewise compact-open. If $(K, V; W)$ is fibrewise compact-open, where $W \subset B$ is open, $V \subset Y_W$ is open and $K \subset W$ is fibrewise compact over W, then $\alpha(K, V; W) = Y_{W-K} \cap V$ which is open since K is closed in W. This completes the proof.

Suppose that $X = B \times T$ for some topological space T. Then each continuous function $\{b\} \times T \to Y_b$ determines, in an obvious way, a continuous function $T \to Y$, so that we have an injection

$$\sigma \colon \text{map}_B(B \times T, Y) \to \text{map}(T, Y),$$

---

† *If X_0 and Y_0 are topological spaces then $\text{map}(X_0, Y_0)$ denotes the set of continuous functions from X_0 to Y_0. If X_0 and Y_0 are pointed topological spaces then $\text{map}^*(X_0, Y_0)$ denotes the subset of continuous pointed functions.*

‡ *The definition of the fibrewise compact-open topology given here is, a priori, somewhat stronger than the corresponding definition in [32].*

where map(T, Y) denotes the ordinary mapping-space with compact-open topology. The image of σ is precisely the set of continuous functions $T \to Y$ of which the images are contained in a single fibre. I assert that σ is an embedding of map$_B(B \times T, Y)$, with the fibrewise compact-open topology, in map(T, Y), with the compact-open topology.

For let (C, V) be a compact-open subset of map(T, Y), so that C is compact in T and V is open in Y. Then

$$\sigma^{-1}(C, V) = (B \times C, V; B),$$

where $B \times C$ is fibrewise compact over B. This shows that σ is continuous.

To show that σ is an embedding it is sufficient to consider the case of a fibrewise compact-open set $(K, V; W)$, where W is open, $K \subset W \times T$ is fibrewise compact over W, and $V \subset Y_W$ is open. We have to show that $(K, V; W)$ is the inverse image, under σ, of an open set of map(T, Y). But V is open in Y and $K_b = \{b\} \overset{h}{\times} C_b$ where $C_b \subset T$ is compact, for each point $b \in W$. Therefore $(K, V; W)$ is the union of the inverse images $\sigma^{-1}(C_b, V)$ $(b \in W)$. Since (C_b, V) is open for each b so is the union, and since $(K, V; W)$ is the inverse image of the union this proves the result.

Clearly if $B' \subset B$ then map$_B(X, Y)|B'$ can be identified with map$_{B'}(X', Y')$, as a fibrewise set over B', where $X' = X_{B'}$, $Y' = Y_{B'}$. Now the restriction $(X_{B'} \cap K, Y_{B'} \cap V; W \cap B')$ to B' of a fibrewise compact-open set $(K, V; W)$ of map$_B(X, Y)$ is a fibrewise compact-open set of map$_{B'}(X', Y')$, hence the identity function

$$\text{map}_{B'}(X', Y') \to \text{map}_B(X, Y)|B'$$

is a continuous bijection. When B' is open in B the function is an equivalence, but this is not true generally.

In particular the induced topology on the fibre map(X_b, Y_b) of map$_B(X, Y)$ over the point b of B may be coarser than the compact-open topology. An exception is when X is fibrewise discrete since in that case a compact subset C of the discrete X_b is necessarily finite, say $C = \{x_1, \dots, x_n\}$. So we can find a neighbourhood W of b and, by using local slices through $x_1, \dots, x_n$, a family $K_1, \dots, K_n$ of subsets of X_W which are fibrewise compact over W and whose union K meets X_b in C. Since K is fibrewise compact over W it follows that the topologies on map(X_b, Y_b) coincide in this case.

Among the fibrewise compact-open sets of map$_B(X, Y)$ a special role is played by those of the form

$$\langle s, V \rangle = (sW, V; W),$$

where W is open in B, where s is a section of X over W, and where V is open in Y. In the case in which X is fibrewise discrete these special fibrewise compact-open sets form a subbasis for the fibrewise compact-open topology, for all fibrewise topological Y. For let $(K, V; W)$ be a fibrewise compact-open set, where W is open in B, where $K \subset X_W$ is fibrewise compact over W, and where V is open in Y. Let $b \in W$ be a given point and assume, to avoid trivialities, that K_b is non-empty. Let $\phi: X_b \to Y_b$ be a continuous function such that $\phi K_b \subset V_b$, i.e. $K_b \subset \phi^{-1} V_b$. We have $K_b = \{x_1, \ldots, x_n\}$, say, since K_b is discrete and compact, therefore finite. Choose a neighbourhood W_i of b, where $W_i \subset W$, and a section $s_i: W_i \to X$ such that $s_i(b) = x_i$, for $i = 1, \ldots, n$. Note that $U_i = s_i W_i$ is open in X, since X is fibrewise discrete. Since K is fibrewise compact over W the subset

$$W_o = \{\beta \in W: \{s_1(\beta), \ldots, s_n(\beta)\} \supset K_\beta\}$$

is open in B. Now $W' = W_0 \cap W_1 \cap \cdots \cap W_n \subset W$ is a neighbourhood of b, and $\langle s_i', V \rangle$ is a neighbourhood of ϕ for $i = 1, \ldots, n$, where $s_i' = s_i | W'$. Since

$$\langle s_i', V \rangle \cap \cdots \cap \langle s_n', V \rangle \subset (K, V; W)$$

this shows that the special fibrewise compact-open sets form a fibrewise subbasis.

The above observation is due to Lever [43] who uses it to establish a generalization of the fibrewise Tychonoff theorem (3.5) including, as a special case

Proposition (9.2). *Let X be fibrewise discrete over B. Then $\mathrm{map}_B(X, Y)$ is fibrewise compact whenever Y is fibrewise compact.*

Let $p: X \to B$ and $q: Y \to B$ be the projections. Given a non-empty open set H of $\mathrm{map}_B(X, Y)$ we have to show that

$$\{b \in B: \mathrm{map}(X_b, Y_b) \subset H\}$$

is open in B. In other words we have to show that for each point b of B such that $\mathrm{map}(X_b, Y_b) \subset H$ there exists a neighbourhood W such that $\mathrm{map}(X_\beta, Y_\beta) \subset H$ whenever $\beta \in W$. We dispose of two special cases first of all.

If X_b is empty then any neighbourhood of the one-point set $\mathrm{map}(X_b, Y_b)$ contains a neighbourhood of the form $\mathrm{map}_W(X_W, Y_W)$, for some neighbourhood W of b. In particular this is true of H, from which this case of the result follows at once.

Next suppose that $\mathrm{map}(X_b, Y_b)$ is empty. Then Y_b is empty and X_b is not. Since qY is closed the complement $B - qY$ is open and contains b.

Since X is fibrewise discrete there exists a neighbourhood $W \subset B - qY$ of b and a neighbourhood V of some point x of X_b which projects homeomorphically onto W. If $\beta \in W$ then $\mathrm{map}(X_\beta, Y_\beta)$ is empty and so contained in H, as required.

Now suppose that $\mathrm{map}(X_b, Y_b)$ is non-empty and hence Y_b is non-empty. Since $\mathrm{map}(X_b, Y_b)$ is compact there exists a finite cover of $\mathrm{map}(X_b, Y_b)$ by fibrewise compact-open sets $H_1, \ldots, H_n$, say, such that $\mathrm{map}(X_b, Y_b)$ meets H_i for $i = 1, \ldots, n$. As we have seen, H_i can be expressed as the intersection

$$\langle s_{i1}, V_{i1} \rangle \cap \cdots \cap \langle s_{im_i}, V_{im_i} \rangle$$

of special fibrewise compact-open sets, where s_{ij} is a local section of X at the point b and where V_{ij} is an open set of Y. Since $\mathrm{map}(X_b, Y_b)$ meets H_i we see at once that V_{ij} meets $Y_{W_{ij}}$, where W_{ij} denotes the domain of s_{ij}.

Let $s_1, \ldots, s_m$ be the list of distinct local sections of X from among the list

$$s_{11}, \ldots, s_{1m_1}, \ldots, s_{n1}, \ldots, s_{nm_n}.$$

Without real loss of generality we can assume that each local section has the same neighbourhood N of b as domain. Now for each $i = 1, \ldots, n$, and $k = 1, \ldots, m$ we have

$$H_i \subset \langle s_k, Y \rangle = \mathrm{map}_N(X_N, Y_N).$$

After renumbering the V_{ij}s appropriately we can rewrite the H_is in the form

$$H_1 = \langle s_1, V_{11} \rangle \cap \cdots \cap \langle s_m, V_{1m} \rangle,$$

$$\vdots$$

$$H_n = \langle s_1, V_{n1} \rangle \cap \cdots \cap \langle s_m, V_{nm} \rangle,$$

with the proviso that $V_{ik} = Y$ if there is no original j with $s_k = s_{ij}$.

Consider, for each index k, the covering $\{V_{ik} : i = 1, \ldots, n\}$ of Y_b. For each point $y \in Y_b$ let $V_k(y)$ denote the intersection of the V_{ik} which contain y. Then $V_k(y)$ is a neighbourhood of y since the intersection is finite. Now Y is fibrewise compact and so, by (3.3), there exists a neighbourhood N_k of b such that $\{V_k(y)\}$ also covers Y_{N_k}. I assert that $\mathrm{map}_W(X_W, Y_W) \subset H$, where $W = N_1 \cap \cdots \cap N_m$.

For let $\beta \in W$. If $\mathrm{map}(X_\beta, Y_\beta)$ is empty then $\mathrm{map}(X_\beta, Y_\beta) \subset H$ trivially, so let us assume that $\mathrm{map}(X_\beta, Y_\beta)$ is non-empty. Let $\phi : X_\beta \to Y_\beta$ be continuous. For each index k there exists a point $y \in Y_b$ such that $\phi s_k(\beta) \in V_k(y)$. Then $\theta s_k(b) = y$ for some continuous $\theta : X_b \to Y_b$. Now $\theta \in H_i$ for some index i so for each index k we have $\theta s_k(b) \in V_{ik}$ and hence $\theta s_k(\beta) \in V_{ik}$ since $\theta s_k(b) = y$ and $\phi s_k(\beta) \in V_k(y)$. Thus $\phi \in H_i \subset H$, as required.

We have shown that $\text{map}_B(X, Y)$ is fibrewise closed. Moreover, each of the fibres $\text{map}(X_b, Y_b)$ is compact, by the ordinary Tychonoff theorem, since X_b is discrete and Y_b compact. Therefore $\text{map}_B(X, Y)$ is fibrewise compact, as asserted.

It is not generally true that $\text{map}_B(X, Y)$ is fibrewise Hausdorff whenever Y is fibrewise Hausdorff, as the following example given by Lewis [44] shows (see Figure 4). Take $B = I$, with the usual topology, and take $X = Y = I^+$, the topological sum of the point-space $*$ and I, with projection the identity on I and $* \to 0$. The fibre $\{*, 0\}$ of X and Y over 0 admits self-maps (i) the identity and (ii) the constant at 0 which are distinct, and yet, regarded as points of the fibre of $\text{map}_B(X, Y)$ over 0, these self-maps do not admit disjoint neighbourhoods, in the fibrewise compact-open topology. However, we are able to prove

Proposition (9.3). *Let X be locally sliceable over B. Then $\text{map}_B(X, Y)$ is fibrewise Hausdorff whenever Y is fibrewise Hausdorff.*

For let $\phi, \psi \colon X_b \to Y_b$ be distinct maps, where $b \in B$. Then $\phi(x) \neq \psi(x)$ for some point x of X_b. Since X is locally sliceable there exists a neighbourhood W of b and a section $s \colon W \to X_W$ such that $s(b) = x$. If Y is fibrewise Hausdorff there exist disjoint neighbourhoods U, V of $\phi(x)$, $\psi(x)$, respectively, in Y. Then $(sW, U; W)$, $(sW, V; W)$ are disjoint neighbourhoods of ϕ, ψ, respectively.

Figure 4

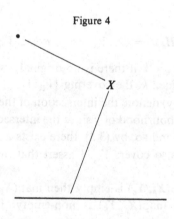

B

Let X, Y and Z be fibrewise topological over B. Precomposition with a continuous fibrewise function $\theta: X \to Y$ determines a continuous fibrewise function

$$\theta^*: \operatorname{map}_B(Y, Z) \to \operatorname{map}_B(X, Z),$$

while postcomposition with a continuous fibrewise function $\phi: Y \to Z$ determines a continuous fibrewise function

$$\phi_*: \operatorname{map}_B(X, Y) \to \operatorname{map}_B(X, Z).$$

Proposition (9.4). *Let $\theta: X \to Y$ be a proper fibrewise surjection, where X and Y are fibrewise topological over B. Then the fibrewise function $\theta^*: \operatorname{map}_B(Y, Z) \to \operatorname{map}_B(X, Z)$ is an embedding for all fibrewise topological Z.*

For let $W \subset B$ be open. If $K \subset Y_W$ is fibrewise compact over W then so is $\theta^{-1}K \subset X_W$. Since θ^* is injective we have $(K, V; W) = \theta^{*-1}(\theta^{-1}K, V; W)$ for each open set V of Z_W. Since $(\theta^{-1}K, V; W)$ is open this proves θ^* is an embedding.

Proposition (9.5). *Let $\phi: Y \to Z$ be a fibrewise embedding, where Y and Z are fibrewise topological over B. Then the fibrewise function*

$$\phi_*: \operatorname{map}_B(X, Y) \to \operatorname{map}_B(X, Z)$$

is an embedding for all fibrewise topological spaces X. If, further, ϕ is closed then ϕ_* is closed provided X is locally sliceable.

The first assertion is almost obvious since if $(K, U; W)$ is a fibrewise compact-open set of $\operatorname{map}_B(X, Y)$ then $U = \phi^{-1}V$ for some open set V of Z_W and so $(K, U; W)$ is the inverse image under ϕ_* of the fibrewise compact-open set $(K, V; W)$ of $\operatorname{map}_B(X, Z)$. To prove the second assertion, let $\alpha: X_b \to Z_b$ $(b \in B)$ belong to the complement of $\phi_* \operatorname{map}_B(X, Y)$ in $\operatorname{map}_B(X, Z)$. Then $\alpha(x) \in U$, for some $x \in X_b$, where $U = Z - \phi Y$ is open. If X is locally sliceable then for some neighbourhood W of b there exists a section $s: W \to X_W$ such that $s(b) = x$. Then sW is fibrewise compact over W and $(sW, U; W)$ is a fibrewise compact-open neighbourhood of α which does not meet $\phi_* \operatorname{map}_B(X, Y)$. This completes the proof.

Proposition (9.6). *Let $\{X_r\}$ be a family of fibrewise topological spaces over B, and let $\sigma_r: X_r \to \coprod_B X_r$ be the standard insertion. Then the fibrewise* *function*

$$\text{map}_B(\textstyle\coprod_B X_r, Y) \to \textstyle\prod_B \text{map}_B(X_r, Y),$$

given by σ_r^* in the rth factor, is a fibrewise topological equivalence, for all fibrewise topological Y.

This follows at once from the definition of the fibrewise compact-open topology since for $W \subset B$ the subsets of $(\coprod_B X_r)_W$ which are fibrewise compact over W are precisely the sums $\coprod_B K_r$ where $K_r \subset (X_r)_W$ is fibrewise compact over W.

Proposition (9.7). *Let X, Y and Z be fibrewise topological over B. If the fibrewise function $h \colon X \times_B Y \to Z$ is continuous then so is the fibrewise function $\hat{h} \colon X \to \text{map}_B(Y, Z)$, where*

$$\hat{h}(x)(y) = h(x, y) \quad (x \in X_b, y \in Y_b, b \in B).$$

For let $(K, V; W)$ be a fibrewise compact-open set of $\text{map}_B(Y, Z)$ and let $x \in \hat{h}^{-1}(K, V; W)$, where $x \in X_b$. Then the trace of $h^{-1}V$ on $X_W \times_W K$ is a neighbourhood of the inverse image $\{x\} \times K_b$ of x under the projection $X_W \times_W K \to X_W \times_W W = X_W$. Since the projection is closed there exists a neighbourhood $U \subset X_W$ of x such that $U \times_W K \subset h^{-1}V$. Then $U \subset \hat{h}^{-1}(K, V; W)$. Thus $\hat{h}$ is continuous, as asserted.

Proposition (9.8). *Let X be fibrewise regular over B. Let Y be fibrewise topological with fibrewise neighbourhood subbasis Γ. Then a fibrewise neighbourhood subbasis for the fibrewise compact-open topology of $\text{map}_B(X, Y)$ consists of the family of fibrewise compact-open subsets $(K, V; W)$, where $W \subset B$ is open, where $K \subset X_W$ is fibrewise compact over W, and where $V \in \Gamma_W$, the trace of Γ on Y_W.*

Note first of all that if $W \subset B$, $K \subset X_W$ and $V_j \subset Y_W$ $(j \in J)$ then

$$\left(K, \bigcap_{j \in J} V_j; W \right) = \bigcap_{j \in J} (K, V_j; W).$$

Without loss of generality, therefore, we may assume that Γ is a fibrewise neighbourhood basis. So let $(K, V; W)$ be an arbitrary fibrewise compact-open neighbourhood of a given $\phi \colon X_b \to Y_b$ $(b \in B)$. Thus W is open in B, $K \subset X_W$ is fibrewise compact over W, $V \subset Y_W$ is open, and $\phi K_b \subset V_b$, so that $K_b \subset \phi^{-1}V_b$. Now V is the union of a family $\{V_j \colon j \in J\}$ of members of Γ_W. For each index j there exists an open $N_j \subset X_W$ such that $\phi^{-1}(V_j \cap Y_b) = N_j \cap X_b$. Since K is fibrewise compact over W and since $\{N_j\}$ covers K_b there exists, by (3.3), a neighbourhood $W' \subset W$ of b such that a finite subfamily of $\{N_j\}$, indexed by $j_1, \ldots, j_n$, say, covers $K' = K_{W'}$. Since X is fibrewise regular there exists, by (3.25), a neighbourhood $W'' \subset W'$ of b and an open covering $\{U_j\}$ $(j = j_1, \ldots, j_n)$ of $K'' = K_{W''}$ such that $X_{W''} \cap \bar{U}_j \subset N_j$ for each j. Now K'' is fibrewise compact over W'' and so $K'' \cap \bar{U}_j$ is fibrewise compact over W'', since $X_{W'} \cap \bar{U}_j$ is closed in $X_{W''}$. Hence

$$(K'' \cap \bar{U}_j, V''_j; W'') \quad (j = j_1, \ldots, j_n),$$

where $V''_j = V_j \cap Y_{W''}$, is a fibrewise compact-open neighbourhood of ϕ. Then the intersection

$$\bigcap_{i=1}^{n} (K'' \cap \bar{U}_j, V''_j; W'') \subset (K'' \cap \bar{U}, V; W'')$$

where $U = \bigcup_{i=1}^{n} U_{j_i}$, is a neighbourhood of ϕ contained in the given neighbourhood $(K, V; W)$. Since the V''_j are the traces on $Y_{W''}$ of members of Γ this proves (9.8), and we at once deduce

Proposition (9.9). *Let Y_1, Y_2 be fibrewise topological over B. Let*

$$Y_1 \xleftarrow{\pi_1} Y_1 \times_B Y_2 \xrightarrow{\pi_2} Y_2$$

be the standard projections. Then the fibrewise function

$$\mathrm{map}_B(X, Y_1 \times_B Y_2) \to \mathrm{map}_B(X, Y_1) \times_B \mathrm{map}_B(X, Y_2)$$

given by π_{1*} in the first factor and by π_{2*} in the second, is a fibrewise topological equivalence for all fibrewise regular X.

Proposition (9.10). *Let X and Y be fibrewise regular over B. Then for all fibrewise topological Z the fibrewise compact-open neighbourhood filters of $\mathrm{map}_B(X \times_B Y, Z)$ are generated by fibrewise compact-open neighbourhoods of the form $(E \times_W F, V; W)$, where $W \subset B$ and $V \subset Z_W$ are open, and where $E \subset X_W$ and $F \subset Y_W$ are fibrewise compact over W.*

First consider a fibrewise compact subspace K of the fibrewise topological product $X \times_B Y$. Since X and Y are fibrewise regular, so is $X \times_B Y$. Let N be a neighbourhood of the compact K_b in $X \times_B Y$, where $b \in B$. Then there exists, for each point $\xi = (x, y)$ of K_b, a neighbourhood W_ξ of b in B and neighbourhoods U_ξ of x in X_{W_ξ}, V_ξ of y in Y_{W_ξ}, such that the closure

$$(X_{W_\xi} \times_{W_\xi} Y_{W_\xi}) \cap (\bar{U}_\xi \times_{W_\xi} \bar{V}_\xi)$$

of $U_\xi \times_{W} V_\xi$ in $X_{W_\xi} \times_{W_\xi} Y_{W_\xi}$ is contained in N. Since K is fibrewise compact over B, moreover, there exists a neighbourhood W of b such that a finite subfamily of $\{U_\xi \times_{W_\xi} V_\xi\}$, indexed by $\xi_1, \ldots, \xi_n$, say, covers K_W. Replacing W by $W' = W \cap W_{\xi_1} \cap \cdots \cap W_{\xi_n}$ and U_{ξ_i}, V_{ξ_i} by $U'_{\xi_i} = X_{W'} \cap U_{\xi_i}$, $V'_{\xi_i} = Y_{W'} \cap V_{\xi_i}$, we obtain the finite open covering $(U'_{\xi_i} \times_{W'} V'_{\xi_i})$ of $K' = K_{W'}$, with

$$(X_{W'} \times_{W'} Y_{W'}) \cap (\bar{U}'_{\xi_i} \times_{W'} \bar{V}'_{\xi_i}) \subset N.$$

Now let L', M' be the projections of K' in $X_{W'}$, $Y_{W'}$, respectively, and write

$$E_i = L' \cap \bar{U}'_{\xi_i}, \qquad F_i = M' \cap \bar{V}'_{\xi_i} \quad (i = 1, \ldots, n).$$

Then E_i and F_i are fibrewise compact over W', since L' and M' are fibrewise compact over W', and $K_{W'}$ is covered by $\{E_i \times_{W'} F_i\}$.

So let $(K, V; W)$ be a fibrewise compact-open neighbourhood of $\phi \colon X_b \times Y_b \to Z_b$, where K, W are as above and $V \subset Z_W$ is open. Then $\phi K_b \subset V_b$ and so $\phi^{-1} V_b = U_b$ for some open $U \subset X_W \times_W Y_W$. Hence, by the above, there exists a neighbourhood $W' \subset W$ of b and fibrewise compact subsets $E_i \subset X_{W'}$, $F_i \subset Y_{W'}$ $(i = 1, \ldots, n)$ over W' such that each fibrewise compact-open set $(E_i \times_{W'} F_i, V'; W')$ is a neighbourhood of ϕ, and such that the family $\{E_i \times_{W'} F_i\}$ covers $K' = K_{W'}$. Then

$$\bigcap_{i=1}^{n} (E_i \times_{W'} F_i, V'; W') = \left(\bigcup_{i=1}^{n} (E_i \times_{W'} F_i), V'; W' \right)$$

is a neighbourhood of ϕ contained in the given neighbourhood $(K, V; W)$. This proves (9.10) and we at once deduce

Proposition (9.11). *Let X_i $(i = 1, 2)$ be fibrewise regular over B and let Y_i $(i = 1, 2)$ be fibrewise topological over B. Then the fibrewise function*

$$\text{map}_B(X_1, Y_1) \times_B \text{map}_B(X_2, Y_2) \to \text{map}_B(X_1 \times_B X_2, Y_1 \times_B Y_2),$$

given by the fibrewise product functor, is an embedding.

Fibrewise composition† determines a fibrewise function

$$\text{map}_B(Y, Z) \times_B \text{map}_B(X, Y) \to \text{map}_B(X, Z),$$

for all fibrewise topological X, Y, Z. In particular (taking $X = B$ and replacing Y, Z by X, Y, respectively) fibrewise evaluation determines a

---

† By fibrewise composition here I mean composition in each fibre. A similar remark applies to the terms of fibrewise evaluation, etc.; used later.

fibrewise function

$$map_B(X, Y) \times_B X \to Y.$$

Proposition (9.12). *Let Y be fibrewise locally compact and fibrewise regular over B. Then the fibrewise function*

$$map_B(Y, Z) \times_B map_B(X, Y) \to map_B(X, Z),$$

given by fibrewise composition, is continuous for all fibrewise topological X and Z.

Corollary (9.13). *Let X be fibrewise locally compact and fibrewise regular over B. Then the fibrewise evaluation function*

$$map_B(X, Y) \times_B X \to Y$$

is continuous for all fibrewise topological Y.

To prove (9.12), let $(K, V; W)$ be a fibrewise compact-open neighbourhood of $\phi\theta \in map_B(X, Z)$, where $\theta: X_b \to Y_b$ and $\phi: Y_b \to Z_b$ $(b \in B)$. Thus W is a neighbourhood of b, $V \subset Z_W$ is open and $K \subset X_W$ is fibrewise compact over W. Now $\theta(X_b \cap K) \subset \phi^{-1}(Z_b \cap V) = Y_b \cap U$ for some open $U \subset Y_W$. Since Y is fibrewise locally compact and fibrewise regular there exists, by (3.14), a neighbourhood $W' \subset W$ of b and a neighbourhood N of $\theta(X_b \cap K)$ in $Y_{W'}$, such that the closure $Y_{W'} \cap \bar{N}$ of N in $Y_{W'}$ is fibrewise compact over W' and contained in U. The fibrewise composition function sends $(X_{W'} \cap \bar{N}, V'; W') \times_{W'} (K', N; W')$ into $(K', V'; W')$ where $K' = X_{W'} \cap K$ and $V' = Z_{W'} \cap V$. Since $\theta \in (K', N; W')$ and $\phi \in (X_{W'} \cap \bar{N}, V'; W')$ this proves (9.12) and hence (9.13).

Corollary (9.14). *Let X, Y and Z be fibrewise topological over B, with Y fibrewise locally compact and fibrewise regular. Let $\hat{h}: X \to map_B(Y, Z)$ be a continuous fibrewise function. Then the fibrewise function $h: X \times_B Y \to Z$ is continuous, where*

$$h(x, y) = \hat{h}(x)(y) \quad (x \in X_b, y \in Y_b, b \in B).$$

This constitutes a converse of (9.7), subject to the restriction on Y. It is usual to refer to h as the *adjoint* of $\hat{h}$, and to $\hat{h}$ as the *adjoint* of h, when the functions are related as above. For the proof of (9.14) it is only necessary to observe that h may be expressed as the composition

$$X \times_B Y \to map_B(Y, Z) \times_B Y \to Z$$

of $\hat{h} \times id_Y$ and the fibrewise evaluation function.

By way of application let us use (9.14) to give an alternative proof of (5.3). This concerns a fibrewise quotient map $\phi \colon X \to Y$, where X and Y are fibrewise topological over B. Let T and Z be fibrewise topological spaces and let f and g be fibrewise functions making the diagram on the left below commutative.

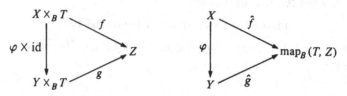

Suppose that f is continuous. Then the adjoints $\hat{f}$ and $\hat{g}$ are defined, as on the right, and $\hat{f}$ is continuous by (9.7). Therefore $\hat{g}$ is continuous, since ϕ is a fibrewise quotient map. If we now assume that T is fibrewise locally compact and fibrewise regular then (9.14) applies and we obtain that g is continuous. Under this further assumption, therefore, we see that $\phi \times \mathrm{id}$ is a fibrewise quotient map, as asserted in (5.3).

We are now ready to establish the exponential law, which justifies the use of the term adjoint in this context.

Proposition (9.15). *Let X, Y and Z be fibrewise topological over B and let*

$$\xi \colon \mathrm{map}_B(X \times_B Y, Z) \to \mathrm{map}_B(X, \mathrm{map}_B(Y, Z))$$

be the fibrewise injection defined by taking adjoints in each fibre. If X is fibrewise regular then ξ is continuous. If both X and Y are fibrewise regular then ξ is an open embedding. If, further, Y is fibrewise locally compact then ξ is fibrewise topological equivalence.

By (9.10), when X is fibrewise regular the fibrewise neighbourhood filters of $\mathrm{map}_B(X, \mathrm{map}_B(Y, Z))$ are generated by fibrewise compact-open neighbourhoods of the form $(E, (F, V; W); W)$, where $W \subset B$ and $V \subset Z_W$ are open, and where $E \subset X_W$ and $F \subset Y_W$ are fibrewise compact over W. The inverse image of $(E, (F, V; W); W)$ is $(E \times_B F, V; W)$, which is also fibrewise compact-open and so ξ is continuous.

By (9.10) when X and Y are fibrewise regular the fibrewise neighbourhood filters of $\mathrm{map}_B(X \times_B Y, Z)$ are generated by fibrewise compact-open neighbourhoods of the form $(E \times_W F, V; W)$, where $W \subset B$ and $V \subset Z_W$ are open and where $E \subset X_W$ and $F \subset Y_W$ are fibrewise compact over W. The direct image of $(E \times_W F, V; W)$ is $(E, (F, V; W); W)$, which is also fibrewise compact-open and so ξ is an open embedding. The final assertion follows at once from (9.14).

We now turn to consideration of the pointed version of the fibrewise mapping-space. As we shall see this constitutes an adjoint of the fibrewise smash product, under certain conditions.

If X and Y are fibrewise pointed topological spaces over B we denote by $\mathrm{map}_B^B(X, Y)$ the subspace $\coprod_{b \in B} \mathrm{map}^*(X_b, Y_b)$ of $\mathrm{map}_B(X, Y)$ consisting of pointed functions on the fibres, where the basepoints in the fibres are determined by the sections in the usual way. We regard $\mathrm{map}_B^B(X, Y)$ as a fibrewise pointed topological space over B, the section being that which sends each point $b \in B$ into the pointed function $X_b \to Y_b$ to the basepoint, and refer to it as the *fibrewise pointed mapping-space*. The section is closed, by (9.5), if X itself has a closed section and is locally sliceable, for arbitrary fibrewise pointed Y. If, further, Y is fibrewise Hausdorff then so is $\mathrm{map}_B^B(X, Y)$.

The next four results are straightforward consequences of the corresponding results in the non-pointed case.

Proposition (9.16). *Let X and Y be fibrewise pointed topological spaces over B. Then $\mathrm{map}_B^B(X, B)$ and $\mathrm{map}_B^B(B, Y)$ are equivalent to B, as fibrewise pointed topological spaces.*

Let X, Y and Z be fibrewise pointed topological spaces over B. Precomposition with a continuous fibrewise pointed function $\theta: X \to Y$ determines a continuous fibrewise pointed function

$$\theta^*: \mathrm{map}_B^B(Y, Z) \to \mathrm{map}_B^B(X, Z)$$

while postcomposition with a continuous fibrewise pointed function $\phi: Y \to Z$ determines a continuous fibrewise pointed function

$$\phi_*: \mathrm{map}_B^B(X, Y) \to \mathrm{map}_B^B(X, Z).$$

Proposition (9.17). *Let $\theta: X \to Y$ be a proper fibrewise pointed surjection, where X and Y are fibrewise pointed topological spaces over B. Then the fibrewise pointed function*

$$\theta^*: \mathrm{map}_B^B(Y, Z) \to \mathrm{map}_B^B(X, Z)$$

is an embedding for all fibrewise pointed topological Z.

Proposition (9.18). *Let $\phi: Y \to Z$ be a fibrewise pointed embedding, where Y and Z are fibrewise pointed topological spaces over B. Then the fibrewise pointed function*

$$\phi_*: \mathrm{map}_B^B(X, Y) \to \mathrm{map}_B^B(X, Z)$$

is an embedding for all fibrewise pointed topological X. If, further, ϕ is closed then ϕ_* is closed provided X is locally sliceable.

Proposition (9.19). *Let X_i ($i = 1, 2$) be a fibrewise pointed topological space over B with closed section and let*

$$X_1 \xrightarrow{\sigma_1} X_1 \vee_B X_2 \xleftarrow{\sigma_2} X_2$$

be the standard insertions. Then the fibrewise function

$$\mathrm{map}_B^B(X_1 \vee_B X_2, Y) \to \mathrm{map}_B^B(X_1, Y) \times_B \mathrm{map}_B^B(X_2, Y),$$

given by σ_1^* in the first factor and by σ_2^* in the second, is an equivalence of fibrewise pointed topological spaces, for all fibrewise pointed topological Y.

Obviously the fibrewise function in question is a continuous bijection. Moreover the fibrewise quotient function

$$X_1 +_B X_2 \to X_1 \vee_B X_2$$

is proper, as we have seen. Hence and from (9.4) the result follows.

Proposition (9.20). *Let X, Y and Z be fibrewise pointed topological spaces over B. If the fibrewise pointed function $h: X \wedge_B Y \to Z$ is continuous then so is the fibrewise pointed function $\hat{h}: X \to \mathrm{map}_B^B(Y, Z)$, where*

$$\hat{h}(x)(y) = h(x, y) \qquad (x \in X_b, y \in Y_b, b \in B).$$

This follows at once from (9.7). Similarly (9.9) implies

Proposition (9.21). *Let Y_i ($i = 1, 2$) be a fibrewise pointed topological space over B and let*

$$Y_1 \xleftarrow{\pi_1} Y_1 \times_B Y_2 \xrightarrow{\pi_2} Y_2$$

be the standard projections. Then the fibrewise function

$$\mathrm{map}_B^B(X, Y_1 \times_B Y_2) \to \mathrm{map}_B^B(X, Y_1) \times_B \mathrm{map}_B^B(X, Y_2),$$

given by π_{1*} in the first factor and by π_{2*} in the second, is an equivalence of fibrewise pointed topological spaces, for all fibrewise regular fibrewise pointed topological X.

Proposition (9.22). *Let X_i be a fibrewise compact and fibrewise regular fibrewise pointed topological space and let Y_i be a fibrewise pointed topo-*

logical space over B ($i = 1, 2$). *Then the fibrewise pointed injection*

$$\text{map}_B^B(X_1, Y_1) \wedge_B \text{map}_B^B(X_2, Y_2) \to \text{map}_B^B(X_1 \wedge_B X_2, Y_1 \wedge_B Y_2)$$

given by the fibrewise smash product is continuous.

This can be established by consideration of the diagram shown below, where ρ is the generic fibrewise quotient map, where ξ is given by the fibrewise topological product, and where η is given by the fibrewise smash product.

$$
\begin{array}{ccc}
\text{map}_B^B(X_1, Y_1) \times_B \text{map}_B^B(X_2, Y_2) & \xrightarrow{\;\;\xi\;\;} & \text{map}_B^B(X_1 \times_B X_2, Y_1 \times_B Y_2) \\
\Big\downarrow{\scriptstyle\rho} & & \Big\downarrow{\scriptstyle\rho_*} \\
& & \text{map}_B^B(X_1 \times_B X_2, Y_1 \wedge_B Y_2) \\
& & \Big\downarrow{\scriptstyle\rho^*} \\
\text{map}_B^B(X_1, Y_1) \wedge_B \text{map}_B^B(X_2, Y_2) & \xrightarrow[\;\;\eta\;\;]{} & \text{map}_B^B(X_1 \wedge_B X_2, Y_1 \wedge_B Y_2)
\end{array}
$$

Now ξ is continuous, by (9.11), while ρ_* and ρ^* on the right are continuous also. Thus $\rho^* \rho_* \xi$ is continuous and so η is continuous, since ρ is a fibrewise quotient map.

In the fibrewise pointed theory fibrewise composition determines a fibrewise pointed function

$$\text{map}_B^B(Y, Z) \wedge_B \text{map}_B^B(X, Y) \to \text{map}_B^B(X, Z)$$

for all fibrewise pointed topological spaces X, Y, Z over B. In particular (take $X = B \times I$ and replace Y, Z by X, Y, respectively) fibrewise evaluation determines a fibrewise pointed function

$$\text{map}_B^B(X, Y) \wedge_B X \to Y.$$

Proposition (9.23). *Let Y be a fibrewise locally compact and fibrewise regular fibrewise pointed topological space over B. Then the fibrewise composition function*

$$\text{map}_B^B(Y, Z) \wedge_B \text{map}_B^B(X, Y) \to \text{map}_B^B(X, Z)$$

is continuous for all fibrewise pointed topological spaces X and Z.

Corollary (9.24). *Let X be a fibrewise locally compact and fibrewise regular fibrewise pointed topological space over B. Then the fibrewise* *evaluation function*

$$\mathrm{map}_B^B(X, Y) \wedge_B X \to Y$$

is continuous for all fibrewise pointed topological spaces Y.

Corollary (9.25). *Let $\hat{h}: X \to \mathrm{map}_B^B(Y, Z)$ be a continuous fibrewise pointed function, where X, Y, Z are fibrewise pointed topological spaces over B with Y fibrewise locally compact and fibrewise regular. Then the fibrewise pointed function $h: X \wedge_B Y \to Z$ is continuous, where*

$$h(x, y) = \hat{h}(x)(y) \quad (x \in X_b, \, y \in Y_b, \, b \in B).$$

All these results, and the next, follow from the corresponding results of the non-pointed theory. We may refer to h as the *adjoint* of $\hat{h}$, and to $\hat{h}$ as the *adjoint* of h, when they are related as above.

Proposition (9.26). *Let X, Y and Z be fibrewise pointed topological spaces over B, with X and Y fibrewise compact and fibrewise regular. Then the fibrewise pointed function*

$$\mathrm{map}_B^B(X \wedge_B Y, Z) \to \mathrm{map}_B^B(X, \mathrm{map}_B^B(Y, Z)),$$

obtained by taking adjoints, is an equivalence of fibrewise pointed topological spaces.

10. Fibrewise compactly-generated spaces

As a further illustration of the ideas discussed earlier we present a fibrewise version of the first part of the well-known article by Steenrod [58] about compactly-generated spaces. In that article, which is influenced by earlier work of Brown [10], the spaces concerned are required to satisfy the Hausdorff property. In our version we shall, of course, require the fibrewise Hausdorff property. In addition we shall also require the existence of local slices, at certain stages in the exposition, but this is not necessary at the outset.

We begin with a few definitions and results which play an auxiliary role in what follows.

Definition (10.1). *Let X be fibrewise topological over B. The subset H of X is* quasi-open *(resp.* quasi-closed*) if the following condition is satisfied: for each point b of B and neighbourhood V of b there exists a neighbourhood*

$W \subset V$ of b such that whenever $K \subset X_W$ is fibrewise compact over W then $H \cap K$ is open (resp. closed) in K.

Obviously the intersection of a finite family of quasi-open sets is quasi-open and the union of a finite family of quasi-closed sets is quasi-closed. It is not, in general, true that infinite unions of quasi-open sets are quasi-open or that infinite intersections of quasi-closed sets are quasi-closed.

Note that $H \subset X$ is quasi-open in X if $X - H$ is quasi-closed in X, and vice versa. Clearly H is quasi-open (resp. quasi-closed) in X if H is open (resp. closed in X), although the converse is generally false.

Proposition (10.2). *Let X be fibrewise locally compact and fibrewise Hausdorff over B. If H is quasi-open (resp. quasi-closed) in X then H is open (resp. closed) in X.*

For suppose that H is quasi-open in X. For each point x of H there exists a neighbourhood V of $b = p(x)$ in B and a neighbourhood U of x in X_V such that the closure $X_V \cap \bar{U}$ of U in X_V is fibrewise compact over V. Since H is quasi-open there exists a neighbourhood $W \subset V$ of b such that $X_W \cap (H \cap \bar{U})$ is open in $X_W \cap \bar{U}$. Then $X_W \cap (H \cap U) = (X_W \cap (H \cap \bar{U})) \cap U$ is open in $X_W \cap U$ and so open in X. This proves the result when H is quasi-open; the result when H is quasi-closed follows by taking complements.

Definition (10.3). *Let X be fibrewise Hausdorff over B. The fibrewise topology of X is fibrewise compactly-generated if every quasi-closed subset of X is closed or, equivalently, if every quasi-open subset of X is open.*

For example, X is fibrewise compactly-generated if X is fibrewise locally compact and fibrewise Hausdorff by (10.2).

Proposition (10.4). *Let X be fibrewise compactly-generated over B. Then $X' = X_{B'}$ is fibrewise compactly-generated over B' for each closed subspace B' of B.*

For let H be quasi-closed in X'. Let b' be a point of B' and let V' be a neighbourhood of b' in B'. There exists a neighbourhood $W' \subset V'$ of b' in B' such that for each fibrewise compact K' over W' the intersection $H \cap K'$ is closed in K'.

I assert that H is quasi-closed in X. For let $b \in B$ and let V be a neighbourhood of b in B. If $b \in B'$ then $V' = B' \cap V$ is a neighbourhood of b in B', so that there exists a neighbourhood $W' \subset V'$ of b in B', as above. Now $W' = B' \cap W$ where $W \subset V$ is a neighbourhood of b in B. If K is fibrewise compact over W then $K' = X' \cap K$ is fibrewise compact over W', since W' is closed in W. So $H \cap K' = H \cap K$ is closed in K' and so closed in K. If $b \notin B'$ then $(B - B') \cap V$ is a neighbourhood of b, contained in V, such that if K is fibrewise compact over $(B - B') \cap V$ then $H \cap K$, being empty, is closed in K. Therefore H is quasi-closed in X and so closed in X, hence closed in X'. Thus X' is fibrewise compactly-generated, as asserted.

Proposition (10.5). *Let X be fibrewise Hausdorff over B. Suppose that $X_j = p^{-1} B_j$ is fibrewise compactly-generated over B_j for each member B_j of a locally finite closed covering of B. Then X is fibrewise compactly-generated over B.*

For let H be quasi-closed in X. I assert that $H_j = X_j \cap H$ is quasi-closed in X_j. For let $b \in B_j$ and let V_j be a neighbourhood of b in B_j. Then $V_j = B_j \cap V$, where V is a neighbourhood of b in B. Since H is quasi-closed there exists a neighbourhood $W \subset V$ such that $H \cap K$ is closed in K whenever $K \subset X_W$ is fibrewise compact over W. Then $H_j \cap K_j$ is closed in K_j whenever $K_j \subset p^{-1} W_j$ is fibrewise compact over $W_j = B_j \cap W$. For K_j is fibrewise compact over W, since W_j is closed in W, and so $H \cap K_j$ is closed in K_j. Therefore $H_j \cap K_j = (H \cap K_j) \cap X_j$ is closed in $K_j \cap X_j = K_j$. Thus H_j is quasi-closed in X_j and so closed in X_j, since X_j is fibrewise compactly-generated over B_j. Therefore H is closed in X, since the B_j form a locally finite closed covering of B. Thus X is fibrewise compactly-generated over B, as asserted.

Proposition (10.6). *Let X be fibrewise compactly-generated over B. Let X' be a fibrewise Hausdorff fibrewise quotient of X. Then X' is fibrewise compactly-generated.*

For consider the natural projection $\pi : X \to X'$. I assert that $\pi^{-1} H' = H$ is quasi-closed in X whenever H' is quasi-closed in X'. For let V be a neighbourhood of a given point b of B. Since H' is quasi-closed there exists a neighbourhood $W \subset V$ of b such that $H' \cap K'$ is closed in K' whenever $K' \subset X'_W$ is fibrewise compact over W. Let $K \subset X_W$ be fibrewise compact over W. Then $\pi K \subset X'_W$ is fibrewise compact over W and so $H' \cap \pi K$ is closed in πK. Therefore $H \cap K = \pi^{-1}(H' \cap \pi K) \cap K$ is closed in $\pi^{-1} \pi K = K$, as required. Hence H is closed in X, since X is fibrewise compactly-generated, and so H' is closed in X', since π is a quotient map. Therefore X' is fibrewise compactly-generated, as asserted.

Next we prove the fibrewise version of a result of Cohen [12].

Proposition (10.7). *Let X be fibrewise Hausdorff over B. Then X is fibrewise compactly-generated if and only if X is fibrewise topologically equivalent to a fibrewise quotient space of a fibrewise locally compact fibrewise Hausdorff space.*

In one direction this follows at once from (10.2) and (10.6). To prove the result in the other direction, suppose that X is fibrewise compactly-generated. Consider the family $\{K_j, W_j\}$ of pairs consisting of an open set W_j of B and a subset K_j of X_{W_j} which is fibrewise compact over W_j. By taking copies, in the usual way, we replace these by pairs $\{K'_j, W_j\}$ of the same description except that the K'_j for different indices j are disjoint. Consider the disjoint union X' of the K'_j, regarded as a fibrewise topological space over B. Each summand K'_j is open and closed in X'. Clearly X' is fibrewise locally compact and is fibrewise Hausdorff since X is fibrewise Hausdorff. By mapping each summand K'_j into K_j and so into X we obtain a continuous fibrewise function $\rho \colon X' \to X$. Then X has the fibrewise quotient topology, for if $H \subset X$ is such that $\rho^{-1}H$ is closed then H is quasi-closed and so closed, since X is fibrewise compactly-generated. This completes the proof.

Definition (10.8). *Let $\phi \colon X \to Y$ be a fibrewise function, where X and Y are fibrewise topological over B. Then ϕ is quasi-continuous if for each point b of B and neighbourhood V of b in B there exists a neighbourhood $W \subset V$ of b such that $\phi \mid K$ is continuous whenever K is fibrewise compact over W.*

Clearly ϕ is quasi-continuous if ϕ is continuous. In the other direction we prove

Proposition (10.9). *Let $\phi \colon X \to Y$ be a fibrewise function, where X is fibrewise compactly-generated and Y is fibrewise Hausdorff over B. If ϕ is quasi-continuous then ϕ is continuous.*

For let H be closed in Y. We show that $\phi^{-1}H$ is quasi-closed in X and so closed in X. For let b be a point of B and let V be a neighbourhood of b in B. Since ϕ is quasi-continuous there exists a neighbourhood $W \subset V$ of b such that $\phi \mid K$ is continuous whenever K is fibrewise compact over

W. Then ϕK is fibrewise compact over W and so ϕK is closed in Y_W. This implies that $H \cap \phi K$ is closed in Y_W and so

$$(\phi \,|\, K)^{-1}(H \cap \phi K) = \phi^{-1}H \cap K$$

is closed in X_W and so closed in K. Therefore $\phi^{-1}H$ is closed in X, since X is fibrewise compactly-generated. Thus ϕ is continuous, as asserted.

We now define a functor k_B from the category of fibrewise Hausdorff spaces over B to the category of fibrewise compactly-generated spaces as follows. If X is a fibrewise Hausdorff space over B we define $k_B X = X'$, the same fibrewise set as X but with the fibrewise topology in which the closed sets of X' are precisely the intersections of the quasi-closed sets of X. Clearly the new fibrewise topology is a refinement of the old one.

To understand the relationship between the two topologies better, let W be a neighbourhood of b in B, let $K \subset X_W$ be a subset in the old topology and let $K' \subset X'_W$ be the same subset in the new topology. Since the new is a refinement of the old it is obvious that K is fibrewise compact over W if K' is fibrewise compact over W. Conversely suppose that K is fibrewise compact over W. If H is closed in K' then $H \cap K = H$ is closed in K. In particular K' is closed in K and so fibrewise compact over W. Thus the subsets which are fibrewise compact over W are the same whichever of the two topologies is used. Moreover if X is already fibrewise compactly-generated then X and X' have the same closed sets, i.e. $X = X'$. In particular $B = B'$ where B, as usual, is regarded as a fibrewise topological space over itself using the identity as projection.

It remains for the functor k_B to be defined on continuous fibrewise functions. So let $\phi \colon X \to Y$ be a fibrewise function, where X and Y are fibrewise Hausdorff over B, and let $\phi' \colon X' \to Y'$ be the same fibrewise function, where $X' = k_B X$ and $Y' = k_B Y$. Suppose that ϕ is quasi-continuous. I assert that ϕ' is quasi-continuous and hence continuous. For let b be a point of B and let V be a neighbourhood of b. Let $W \subset V$ be a neighbourhood of b such that $\phi \,|\, K$ is continuous for all fibrewise compact K over W. Then ϕK is fibrewise compact over W, equivalently $\phi' K'$ is fibrewise compact over W. Thus the fibrewise function $\phi' \,|\, K' \colon K' \to \phi K'$ factors into the composition of $\phi \,|\, K$ and two identity functions $K' \to K \to \phi K \to \phi K'$. Hence $\phi' \,|\, K'$ is continuous and so ϕ' is quasi-continuous, hence continuous. We may therefore define $k_B \phi = \phi'$.

We now turn to the subject of fibrewise topological products. In the first place, if X and Y are fibrewise compactly-generated over B we define their fibrewise compactly-generated fibrewise product to be the result of applying the functor k_B to the ordinary fibrewise topological product. As for notation we continue to use $\times_B$ to denote the ordinary fibrewise topological product but now we use $\times'_B$ to denote the fibrewise compactly-generated fibrewise product, given by

$$X \times'_B Y = k_B(X \times_B Y).$$

It is easy to see that this construction has the necessary formal properties. To begin with, the projections of the fibrewise compactly-generated fibrewise product into its factors are continuous, since the identity function $X \times'_B Y \to X \times_B Y$ is continuous. To establish the Cartesian property, let A be fibrewise compactly-generated and let

$$X \overset{\phi}{\leftarrow} A \overset{\psi}{\to} Y$$

be a cotriad of continuous fibrewise functions. Then ϕ and ψ are components of a unique continuous fibrewise function $\theta: A \to X \times_B Y$. Applying the functor k_B and using the fact that $A' = A$ we obtain the unique continuous fibrewise function $A \to X \times'_B Y$ which, when composed with the projections, gives ϕ and ψ. It follows from this result that the fibrewise compactly-generated fibrewise product satisfies the usual laws of commutativity and associativity. We can extend the construction to any number of factors similarly.

Proposition (10.10). *Let X be a fibrewise locally compact and fibrewise Hausdorff space over B and let Y be a fibrewise compactly-generated space over B. Then the ordinary fibrewise topological product $X \times_B Y$ is fibrewise compactly-generated, and so $X \times'_B Y = X \times_B Y$.*

Since Y is fibrewise compactly-generated there exists, by (10.7), a fibrewise quotient map $\pi: T \to Y$, where T is fibrewise locally compact and fibrewise Hausdorff over B. Since X is fibrewise locally compact and fibrewise Hausdorff so is the fibrewise topological product $X \times_B T$, by (2.11) and (3.15). Moreover

$$\text{id} \times \pi: X \times_B T \to X \times_B Y$$

is a fibrewise quotient map, by (5.3), and so $X \times_B Y$ is fibrewise compactly-generated, by (10.7) again. This proves (10.10).

Proposition (10.11). *Let $\phi: X \to Y$ be a fibrewise quotient map, where X and Y are fibrewise compactly-generated over B. Then the fibrewise product*

$$\phi \times \text{id}: X \times'_B T \to Y \times'_B T$$

is also a fibrewise quotient map, for all fibrewise compactly-generated T.

To prove (10.11), let H be a subset of $Y \times'_B T$ such that the inverse image $(\phi \times \text{id})^{-1} H$ is closed in $X \times'_B T$. Let b be a point of B and let V be a neighbourhood of b in B. Since $Y \times'_B T$ is fibrewise compactly-generated there exists a neighbourhood $W \subset V$ of b such that $H \cap K$ is closed in K whenever $K \subset Y_W$ is fibrewise compact over W. So let $K \subset Y \times'_B T$ be fibrewise compact over W. Let L, M denote the projections of K in Y, T, respectively. Then L, M are fibrewise compact over W, and so $L \times'_W M$ is fibrewise compact over W. If we can show that $H \cap (L \times'_W M)$ is closed in $L \times'_W M$ it will follow that $H \cap K$ is closed in K and then since $Y \times'_B T$ is fibrewise compactly-generated this will establish that H is closed and thereby prove the proposition.

Now since $(\phi \times \text{id})^{-1}(L \times'_W M) = \phi^{-1} L \times'_W M$ is closed in $X_W \times'_W T_W$ it follows that $(\phi \times \text{id})^{-1}(H \cap (L \times'_W M))$ is closed in $\phi^{-1} L \times'_W M$. Substituting (X, Y, T) for $(\phi^{-1} L, L, M)$, we have reduced the proof to the case when Y and T are fibrewise compact over B. In that case we have

$$X \times'_B T = X \times_B T, \qquad Y \times'_B T = Y \times_B T,$$

by (10.10); hence the result follows at once from (5.3) above.

Proposition (10.12). *If X and Y are fibrewise Hausdorff over B then*

$$X' \times'_B Y' = (X \times_B Y)'.$$

Observe first of all that the identity function $X \times'_B Y \to X \times_B Y$ is continuous since the identity functions $X' \to X$ and $Y' \to Y$ are continuous. Hence for each point b of B and neighbourhood W of b the subsets of $X'_W \times_W Y'_W$ which are fibrewise compact over W are also fibrewise compact over W as subsets of $X_W \times_W Y_W$. In fact the converse also holds. For let K be a subset of $X_W \times_W Y_W$ which is fibrewise compact over W. Then the projections $L \subset X_W$ and $M \subset Y_W$ are also fibrewise compact over W. Hence the restriction to $L \times_W M$ of the identity function is bicontinuous. Since $K \subset L \times_W M$ it follows that K is fibrewise compact over W, as a subset of $X'_W \times_W Y'_W$. So $X_W \times_W Y_W$ and $X'_W \times_W Y'_W$ have the same fibrewise compact subsets, as fibrewise topological spaces over W. Consequently it follows from the definition of the functor k_B that their associated fibrewise compactly-generated topologies coincide.

We turn now to the subject of fibrewise mapping-spaces. For fibrewise Hausdorff X and Y over B we recall from (9.3) that $\text{map}_B(X, Y)$ is fibrewise Hausdorff provided X is locally sliceable. Making this further assumption we write

$$\text{map}'_B(X, Y) = k_B \, \text{map}_B(X, Y)$$

and call this the *fibrewise compactly-generated fibrewise mapping-space.*

Our aim is to show that for fibrewise compactly-generated X and Y, with X locally sliceable, this functor has the formal properties of an adjoint to the fibrewise compactly-generated fibrewise product but without the restrictions encountered in ordinary fibrewise topology.

To begin with consider the fibrewise evaluation function

$$\alpha: \text{map}_B(X, Y) \times_B X \to Y,$$

as in §9. The proof of (9.13) shows that for each point b of B and neighbourhood V of b there exists a neighbourhood $W \subset V$ of b such that the restriction of α to subsets of the domain which are fibrewise compact over W is continuous. Therefore the same function

$$k_B(\text{map}_B(X, Y) \times_B X) \to k_B Y$$

is continuous over W. When X is fibrewise compactly-generated the new domain is equivalent, by (10.12), to $\text{map}'_B(X, Y) \times'_B X$. Hence when Y also is fibrewise compactly-generated the fibrewise evaluation function

$$\text{map}'_B(X, Y)^{\iota} \times'_B X \to Y$$

is continuous.

Note that if X is fibrewise compactly-generated then

$$\text{map}_B(X, Y') = \text{map}_B(X, Y)$$

as fibrewise sets, for all fibrewise Hausdorff Y. If further X is locally sliceable then applying the functor k_B to both sides we find that

(10.13) $$\text{map}'_B(X, Y') = \text{map}'_B(X, Y)$$

as fibrewise topological spaces. To establish this it is sufficient to show that for each point b of B and neighbourhood V of b there exists a neighbourhood $W \subset V$ of b such that $\text{map}_B(X, Y')$ and $\text{map}_B(X, Y)$ have the same fibrewise compact sets over W.

In one direction this is easily seen. For since the identity function $Y' \to Y$ is continuous it follows that the identity function

$$\text{map}_B(X, Y') \to \text{map}_B(X, Y)$$

is continuous. So in this direction the assertion is true regardless of the choice of W.

To prove the assertion in the other direction we show that there exists such a neighbourhood W with the following property. Let K be a subset of $\text{map}_B(X, Y)$ which is fibrewise compact over W, and let K^* be the same set retopologized as a subset of $\text{map}_B(X, Y')$. We shall show that each open set N of $(\text{map}_B(X, Y'))_W$ meets K^* in an open set of K. This implies that the inverse correspondence $K \to K^*$ is continuous, whence K^* is fibrewise compact over W.

We choose W, in relation to V, so that the fibrewise evaluation function

$$\alpha: \mathrm{map}_B(X, Y) \times_B X \to Y$$

is continuous on subsets of the domain which are fibrewise compact over W. It obviously suffices to prove the assertion when N is a fibrewise compact-open set $(L, V; W)$, where L is fibrewise compact over W and V is open in Y'. Suppose then that $\phi: X_b \to Y_b$ $(b \in B)$ is a continuous function belonging to the intersection of $(L, V; W)$ with K. Since $K \times_W L$ is fibrewise compact over W it follows that the fibrewise evaluation function $K \times_B L \to Y'$ is continuous over W. Hence the inverse image of V is an open set of $(K \times_B L)_W$. The inverse image is a neighbourhood of $\{\phi\} \times L_b$ and so contains the fibrewise product $U \times_W L$ for some neighbourhood U of ϕ in K, since the projection $K \times_W L \to K$ is closed. It follows that $\phi \in U \subset (L, V; W)$, which proves the assertion and completes the proof of (10.13).

Proposition (10.14). *Let X, Y and Z be fibrewise compactly-generated over B, with X locally sliceable. Then the fibrewise function*

$$\mathrm{map}'_B(X, Y \times'_B Z) \to \mathrm{map}'_B(X, Y) \times'_B \mathrm{map}'_B(X, Z),$$

defined as in (9.9), is an equivalence of fibrewise topological spaces.

We have already shown, in (9.9), that the fibrewise function

$$\mathrm{map}_B(X, Y \times_B Z) \to \mathrm{map}_B(X, Y) \times_B \mathrm{map}_B(X, Z)$$

is an equivalence of fibrewise topological spaces. Apply the functor k_B to both sides. By (10.12) the domain becomes

$$k_B \, \mathrm{map}_B(X, Y \times_B Z) = k_B \, \mathrm{map}_B(X, k_B(Y \times_B Z)) = \mathrm{map}'_B(X, Y \times'_B Z).$$

By (10.12) again the codomain becomes

$$k_B(\mathrm{map}_B(X, Y) \times_B \mathrm{map}_B(X, Z)) = \mathrm{map}'_B(X, Y) \times'_B \mathrm{map}'_B(X, Z),$$

which completes the proof.

Next we establish the exponential law, as in

Proposition (10.15). *Let X, Y and Z be fibrewise compactly-generated over B, with X and Y locally sliceable. Then the fibrewise function*

$$\mathrm{map}'_B(X \times'_B Y, Z) \to \mathrm{map}'_B(X, \mathrm{map}'_B(Y, Z)),$$

defined as in (9.15), is an equivalence of fibrewise topological spaces.

The proof of (10.15) depends on

Lemma (10.16). *Let X, Y and Z be fibrewise compactly-generated over B. Then the adjoint of a continuous fibrewise function $h: X \times'_B Y \to Z$ is a continuous fibrewise function $\hat{h}: X \to \text{map}_B(Y, Z)$.*

For consider a fibrewise compact-open subset $(K, V; W)$ of $\text{map}_B(Y, Z)$, where K is fibrewise compact over W and V is open in Z. Suppose that $\hat{h}(x) \in (K, V; W)$, where $x \in X_b$ for some point b of W, so that $h(\{x\} \times K_b) \subset V_b$. Now $X_W \times_W K \to X_W$ is closed since K is fibrewise compact over W. Also $h^{-1}V$ is quasi-open in $X \times_B Y$ and so there exists a neighbourhood N of x in X_W such that $N \times_W K \subset h^{-1}V$. Then N is contained in $\hat{h}^{-1}(K, V; W)$, and (10.16) follows.

To prove (10.15) we first consider the fibrewise function

$$\mu: \text{map}_B(X \times'_B Y, Z) \to \text{map}_B(X, \text{map}_B(Y, Z)),$$

defined by taking adjoints as in (10.16). We shall show that μ is a fibrewise topological equivalence. Then we shall apply the functor k_B to both sides. The domain of μ becomes $\text{map}'_B(X \times'_B Y, Z)$, by definition. The codomain becomes

$$\text{map}'_B(X, \text{map}_B(Y, Z)) = \text{map}'_B(X, \text{map}'_B(Y, Z)),$$

by (10.13). Hence (10.15) will follow as soon as we have shown that μ is a fibrewise topological equivalence.

To this end we start with the continuity of the fibrewise evaluation function, rearranged as

$$\alpha: X \times'_B \text{map}_B(X \times'_B Y, Z) \times'_B Y \to Z.$$

By (10.16) with X replaced by $X \times'_B \text{map}_B(X \times'_B Y, Z)$, we find that

$$\hat{\alpha}: X \times'_B \text{map}_B(X \times'_B Y, Z) \to \text{map}_B(Y, Z)$$

is continuous. Switching factors and applying (10.16) again, with X replaced by $\text{map}_B(X \times'_B Y, Z)$, Y by X, and Z by $\text{map}_B(Y, Z)$, we find that

$$\hat{\hat{\alpha}}: \text{map}_B(X \times'_B Y, Z) \to \text{map}_B(X, \text{map}_B(Y, Z))$$

is continuous. It is easy to check that $\hat{\hat{\alpha}}$ coincides with μ, in our assertion, so that the continuity of μ is established.

To show that μ has a continuous inverse, consider the fibrewise evaluation functions

$$\alpha: \text{map}_B(X, \text{map}_B(Y, Z)) \times'_B X \to \text{map}_B(Y, Z),$$

$$\alpha: \text{map}_B(Y, Z) \times'_B Y \to Z.$$

Since the fibrewise spaces are fibrewise compactly-generated the composition

$$\alpha(\alpha \times \mathrm{id}) : \mathrm{map}_B(X, \mathrm{map}_B(Y, Z)) \times'_B X \times'_B Y \to Z$$

is continuous. Using the lemma once more, but with X replaced by $\mathrm{map}_B(X, \mathrm{map}_B(Y, Z))$, Y by $X \times_B Y$ and Z by Z, we see that the adjoint

$$\mathrm{map}_B(X, \mathrm{map}_B(Y, Z)) \to \mathrm{map}_B(X \times_B Y, Z)$$

of the compositions is defined and continuous. It is easy to check that the adjoint coincides with the inverse of μ, in our assertion, and so the proof that μ is a fibrewise topological equivalence is complete.

Proposition (10.17). *If X, Y and Z are fibrewise compactly-generated over B, with X and Y locally sliceable, then the fibrewise composition function*

$$\mathrm{map}'_B(Y, Z) \times'_B \mathrm{map}'_B(X, Y) \to \mathrm{map}'_B(X, Z)$$

is continuous.

For as we have seen the fibrewise functions

$$\mathrm{map}'_B(Y, Z) \times'_B \mathrm{map}'_B(X, Y) \times'_B X \xrightarrow{\mathrm{id} \times \alpha} \mathrm{map}'_B(Y, Z) \times'_B Y \xrightarrow{\alpha} Z$$

are continuous, hence their composition $\alpha(\mathrm{id} \times \alpha)$ is continuous. Applying (10.16), with X replaced by $\mathrm{map}'_B(Y, Z) \times'_B \mathrm{map}'_B(X, Y)$, Y by X and Z by Z, and with h replaced by $\alpha(\mathrm{id} \times \alpha)$, we see that the adjoint

$$\mathrm{map}'_B(Y, Z) \times'_B \mathrm{map}'_B(X, Y) \to \mathrm{map}_B(X, Z)$$

of the composition is continuous. Now apply the functor k_B and the result is obtained.

The theory can, of course, be developed further but I hope I have taken it far enough to show how the fibrewise version of the convenient category may be useful, in fibrewise homotopy theory, for example.

11. Naturality and naturalization

We begin this section with some generalities. First let $\lambda : B' \to B$ be a function, where B and B' are sets. We can regard B' as a fibrewise set over B, using λ as projection. Then if X is a fibrewise set over B we can regard the fibrewise product $B' \times_B X$ as a fibrewise set over B', using the first projection. We refer to this as the *fibrewise pull-back* of X with respect to λ and denote it by λ^*X. The function $\mu: \lambda^*X \to X$ given by the second projection will be referred to as the *canonical function* over λ.

If $\phi: X \to Y$ is a fibrewise function, where X and Y are fibrewise sets over B, then the fibrewise pull-back $\lambda^*\phi: \lambda^*X \to \lambda^*Y$ is defined, as a fibrewise function over B', by taking the fibrewise product of the identity on B' and ϕ.

If $\{X_r\}$ is a family of fibrewise sets over B then the fibrewise product $\prod_{B'}(\lambda^*X_r)$ of the fibrewise pull-backs is fibrewise equivalent to the fibrewise pull-back $\lambda^* \prod_B(X_r)$ of the fibrewise product.

In the case in which B' is a subset of B and λ is the inclusion then the fibrewise pull-back λ^*X of a fibrewise set X over B can be identified with the restriction X' of X to B', under fibrewise equivalence, and similarly with fibrewise functions.

Now let $\lambda: B' \to B$ be continuous, where B and B' are topological spaces. We can regard B' as fibrewise topological over B, using λ as projection. Then if X is fibrewise topological over B we can regard the fibrewise topological product $B' \times_B X = \lambda^*X$ as a fibrewise topological space over B', using the first projection. We refer to λ^*X as the *fibrewise topological pull-back* of X with respect to λ. Note that the canonical function $\mu: \lambda^*X \to X$ over λ is continuous, and that $\mu^*\mathcal{N}_x$ provides a fibrewise basis for the neighbourhood b'-filter of any point $x' \in \mu^{-1}(x)$, where $b' = p'(x')$ and $x = \mu(x')$. These procedures are obvious functorial, so that if $\phi: X \to Y$ is a continuous fibrewise function over B then $\lambda^*\phi: \lambda^*X \to \lambda^*Y$ is a continuous fibrewise function over B'.

In the case in which B' is a subspace of B and λ the inclusion then the fibrewise topological pull-back λ^*X of a fibrewise topological space X over B can be identified with the restriction X' of X to B', under a fibrewise topological equivalence, and similarly for continuous fibrewise functions.

Also note that if $\{X_r\}$ is a family of fibrewise topological spaces over B then $\lambda^* \prod_B X_r$ is naturally equivalent to $\prod_{B'} \lambda^*X_r$, as a fibrewise topological space over B', for each topological space B' and continuous function $\lambda: B' \to B$.

In the course of Chapter I we considered various classes of fibrewise topological spaces over a given base, and also various classes of fibrewise functions. Let us say that a class of fibrewise topological spaces, or the fibrewise property P which defines it, is *natural* if fibrewise topological pull-backs of members of that class are also members of that class. Specifically, if X is fibrewise P over B, then λ^*X is fibrewise P over B' for each topological space B' and continuous function $\lambda: B' \to B$ and similarly for fibrewise functions.

Since we are only interested in fibrewise properties which are invariant, up to fibrewise topological equivalence, we see, by taking λ to be an embedding, that if X is fibrewise P over B, where fibrewise P is a natural property, then $X_{B'}$ is fibrewise P over B' for each subspace B' of B, and similarly for fibrewise functions. However, it is not always the case that this restricted form of naturality implies naturality for fibrewise topological pull-backs generally.

Obviously the whole class of fibrewise topological spaces is natural, in the above sense, likewise the class of continuous fibrewise functions. Following the line of development in §1, let us show that the class of open fibrewise functions is also natural. Specifically, let $\phi: X \to Y$ be an open fibrewise function, where X and Y are fibrewise topological over B. Let $\lambda: B' \to B$ be continuous, where B' is topological. Then

$$\text{id} \times \phi: B' \times X \to B' \times Y$$

is open, since id and ϕ are open. We may regard id $\times \phi$ as a fibrewise function over $B' \times B$, in the obvious way, and then restrict to the graph of λ. We obtain that the fibrewise product

$$\text{id} \times \phi: B' \times_B X \to B' \times_B Y$$

is open, in other words the fibrewise pull-back $\lambda^*\phi$ of ϕ is open. This shows that the class of open fibrewise functions is natural and hence, as a special case, that the class of fibrewise open spaces is natural.

If we attempt to extend this argument to closed, rather than open functions, we are blocked at the first step, since id $\times \phi$ is not necessarily closed if ϕ is closed. In fact the class of closed fibrewise functions is not natural, nor is the class of fibrewise closed spaces, as can easily be seen by examples. The class of proper functions, however, has precisely the property needed here and so is natural, by the same argument as we used in the case of open functions. In particular the class of fibrewise compact spaces is natural.

The same is true for the class of fibrewise locally compact spaces. Thus let X be fibrewise locally compact over B, and let $X' = \lambda^*X$ be the fibrewise topological pull-back of X with respect to the continuous function $\lambda: B' \to B$. Let $x' \in X'_{b'}$, where $b' \in B'$, and let V' be a neighbourhood of x' in X'. Without real loss of generality we may assume that $V' = \mu^{-1}V$, where V is a neighbourhood of $x = \mu(x')$ in X and μ is canonical. Since X is fibrewise locally compact over B there exists a neighbourhood W of $b = p(x) = \lambda(b')$ in B and a neighbourhood U of x in X_W such that the closure $X_W \cap \bar{U}$ of U in X_W is fibrewise compact over W and contained in V. Then $U' = \mu^{-1}U$ is a neighbourhood of x' in $X'_{W'}$, where $W' = \lambda^{-1}W$, such that the closure $X'_{W'} \cap \bar{U}'$ of U' in $X'_{W'}$ is fibrewise compact over W' and contained in V', as required.

The class of locally sliceable spaces is also natural. For let X be fibrewise topological over B. Let $\lambda: B' \to B$ be continuous, where B' is a topological space, and write $X' = \lambda^*X$, as a fibrewise topological space over B'. Let $x \in X_b$, where $b \in B$, and let $b' \in \lambda^{-1}(b)$. When X is locally sliceable there exists a neighbourhood W of b and a section $s: W \to X_W$ such that $s(b) = x$. Then $\lambda^{-1}W = W'$ is a neighbourhood of b' and

$$\text{id} \times s: W' \times_W W \to W' \times_W X_W$$

constitutes a section $s': W' \to X_{W'}$ such that $s'(b') = (b', x)$. Similarly the class of locally sectionable spaces is natural. Naturality for the class of fibrewise discrete spaces may also be established directly but it may be easier to deduce it from (1.21) above.

Now let us turn to the fibrewise separation conditions, as in §2. Let X, as before, be fibrewise topological over B and write $X' = \lambda^*X$, the fibrewise topological pull-back with respect to $\lambda: B' \to B$. Let $\xi, \eta \in X_b$, where $b \in B$ and $\xi \neq \eta$, and let $b' \in \lambda^{-1}(b)$. If X is fibrewise Hausdorff there exist neighbourhoods U, V of ξ, η in X which do not intersect. Then λ^*U, λ^*V are neighbourhoods of (b', ξ), (b', η) in λ^*X which do not intersect. If X is fibrewise functionally Hausdorff there exists a neighbourhood W of b and a continuous function $\alpha: X_W \to I$ such that $\alpha(\xi) = 0$ and $\alpha(\eta) = 1$. By composing α with the canonical function $\mu_{W'}: X'_{W'} \to X_W$, where $W' = \lambda^{-1}W$, we obtain a continuous function $\alpha': X_{W'} \to I$ such that $\alpha'(b', \xi) = 0$ and $\alpha'(b', \eta) = 1$, as required. Thus the class of fibrewise Hausdorff spaces and the class of fibrewise functionally Hausdorff spaces are both natural, in our sense.

The situation is similar in the case of regularity. Thus let $x \in X_b$, where $b \in B$, and let $b' \in \lambda^{-1}(b)$. Let λ^*U be a fibrewise basic neighbourhood of (b', x) in X', where U is a neighbourhood of x in X. If X is fibrewise regular then there exists a neighbourhood W of b in B and a closed neighbourhood $X_W \cap \bar{V} \subset U$ of x in X_W. Then $X'_{W'} \cap \lambda^*\bar{V} \subset \lambda^*U$ is a closed neighbourhood of (b', x) in $X'_{W'}$, where $W' = \lambda^{-1}W$. If X is fibrewise completely regular, on the other hand, there exists a neighbourhood W of b and a continuous function $\alpha: X_W \to I$ such that $\alpha(x) = 0$ and such that $\alpha = 1$ away from U_W. By precomposing α with the canonical function $X'_{W'} \to X_W$ we obtain a continuous function $\alpha': X'_{W'} \to I$ such that $\alpha'(b', x) = 0$ and such that $\alpha' = 1$ away from $(\lambda^*U)_{W'}$. Thus the class of fibrewise regular spaces and the class of fibrewise completely regular spaces are both natural, in our sense.

Similar questions arise in the pointed case. Thus let X and Y be fibrewise pointed topological spaces over B with closed sections. The fibrewise coproduct behaves naturally, without restriction, since $X \vee_B Y$ is a proper fibrewise quotient of $X +_B Y$. If X and Y are also fibrewise compact then $X \wedge_B Y$ is a proper fibrewise quotient of $X \times_B Y$, as we have seen in §5, and so the fibrewise smash product behaves naturally with this restriction.

Fibrewise Alexandroff compactification also has satisfactory naturality properties. For let X be fibrewise locally compact and fibrewise Hausdorff over B. As we have already seen, the fibrewise topological pull-back λ^*X is fibrewise locally compact and fibrewise Hausdorff over B' for each topological space B' and continuous function $\lambda: B' \to B$. Hence the fibrewise compactifications of X and λ^*X are defined, as fibrewise pointed topological spaces over B and B', respectively, and there is a continuous bijection

$$\theta: (\lambda^*X)^+_{B'} \to \lambda^*(X^+_B)$$

of fibrewise pointed topological spaces over B', given by the canonical fibrewise function. Now X^+_B is fibrewise Hausdorff over B, hence $\lambda^*(X^+_B)$ is fibrewise Hausdorff over B', and $(\lambda^*X)^+_{B'}$ is fibrewise compact over B'. Therefore θ is an equivalence of fibrewise pointed topological spaces over B', by (3.20). It follows as a special case that $(B \times T)^+_B$ is equivalent to $B \times T^+$, as a fibrewise pointed topological space, for locally compact and Hausdorff T. For example $(B \times \mathbb{R}^n)^+_B$ is equivalent to $B \times (\mathbb{R}^n)^+$ and hence to $B \times S^n$.

The concept of naturality can obviously be extended to endofunctors of the category of fibrewise topological spaces over a given base. The fibrewise topological product and coproduct are natural, without restriction, but the fibrewise mapping-space is only natural if some restriction is imposed. Thus let X and Y be fibrewise topological over B. Let $\lambda: B' \to B$ be continuous, where B' is a topological space. For formal reasons we always have a continuous fibrewise bijection

$$\lambda_\#: \mathrm{map}_{B'}(\lambda^*X, \lambda^*Y) \to \lambda^* \mathrm{map}_B(X, Y)$$

of fibrewise topological spaces over B'. Suppose that X is fibrewise locally compact and fibrewise regular. By (9.13) the fibrewise evaluation function

$$\mathrm{map}_B(X, Y) \times_B X \to Y$$

is continuous, hence so is the fibrewise pull-back

$$\lambda^*(\mathrm{map}_B(X, Y) \times_B X) \to \lambda^*Y.$$

Now the domain here is canonically equivalent to $\lambda^* \mathrm{map}_B(X, Y) \times_{B'} \lambda^*X$, and the adjoint of the pull-back of the fibrewise evaluation function is just the inverse $\lambda_{\#}^{-1}$ of the continuous bijection $\lambda_{\#}$. Since the adjoint is continuous, by (9.7), we conclude that $\lambda_{\#}$ is a fibrewise topological equivalence. Thus the fibrewise mapping-space $\mathrm{map}_B(X, Y)$ behaves naturally for fibrewise locally compact and fibrewise regular X and arbitrary fibrewise topological Y.

It follows as a special case that $\mathrm{map}_B(B \times X_0, B \times Y_0)$ is equivalent to $B \times \mathrm{map}(X_0, Y_0)$, as a fibrewise topological space, for locally compact and regular X_0 and topological Y_0.

Similarly the fibrewise pointed mapping-space $\mathrm{map}_B^B(X, Y)$ behaves naturally for fibrewise locally compact and fibrewise regular fibrewise pointed X and arbitrary fibrewise pointed topological Y. In particular $\mathrm{map}_B^B(B \times X_0, B \times Y_0)$ is equivalent to $B \times \mathrm{map}^*(X_0, Y_0)$, as a fibrewise pointed topological space, for all locally compact and regular pointed X_0 and arbitrary pointed topological Y_0.

Of course any property of fibrewise topological spaces can be *naturalized* by the following obvious procedure. Let P_B be a property of fibrewise topological spaces over B, for any topological B. Let us say that the fibrewise topological space X over B has property $P_B^\natural$ if the fibrewise topological pull-back λ^*X has property $P_{B'}$ for each topological space B' and continuous function $\lambda: B' \to B$. For example the class of fibrewise compact spaces arises from the class of fibrewise closed spaces by this process of naturalization.

Exercises

1. Let X be fibrewise Hausdorff over B, and let $\{X_1, X_2\}$ be an open covering of X. Suppose that X_i can be fibrewise embedded as a closed subspace of $B \times \mathbb{R}^{n_i}$ $(i = 1, 2)$. Show that X can be fibrewise embedded as a subspace of $B \times \mathbb{R}^n$ for $n = n_1 + n_2 + 2$.

2. Let X be fibrewise compact and fibrewise regular over B. Show that the topological space (with compact-open topology) of sections of $\mathrm{map}_B(X, Y)$, for all fibrewise topological Y, is equivalent to the topological space of continuous fibrewise functions $X \to Y$ (also with compact-open topology).

3. Show that if X is fibrewise topological and Y is fibrewise pointed topological over B then $\mathrm{map}_B(X, Y)$ itself can be regarded as a fibrewise pointed topological space. Moreover $\mathrm{map}_B(X, Y)$ is then equivalent to $\mathrm{map}_B^B(X/_B \varnothing, Y)$.

4. Regard $B \times I$ as a fibrewise pointed topological space over B with sections given by $b \to (b, 0)$ $(b \in B)$. Show that each fibrewise pointed topological space Y is naturally equivalent to the fibrewise pointed mapping-space $\mathrm{map}_B^B(B \times I, Y)$ through evaluation at the points $(b, 1)$ $(b \in B)$.

5. Let X be locally sliceable and fibrewise compactly-generated over B, and let Y be a fibrewise Hausdorff fibrewise quotient of X. Show that for any fibrewise compactly-generated space Z the fibrewise function

$$\rho^*: \mathrm{map}'_B(Y, Z) \to \mathrm{map}'_B(X, Z)$$

is an embedding, where ρ^* is given by precomposition with the projection ρ.

6. Let $\phi: X' \to X$ be a closed fibrewise embedding, where X and X' are fibrewise topological over B. Show that X' is fibrewise compactly-generated if X is fibrewise compactly-generated.

7. Let X be fibrewise compactly-generated over B. Show that X is then fibrewise compactly-generated over B' for each topological space B' and closed embedding $\beta: B \to B'$.

III
Fibrewise uniform spaces

12. Fibrewise uniform structures

The classical theory of uniform spaces, as in [62], is derived from the theory of metric spaces by abstraction. It does not apply to the Minkowski space of relativity, for example, where the distance is given by an indefinite bilinear form, and is not a metric at all in the topological sense. However, at any given time the metric on the space of events at that time is euclidean. This suggests that we regard Minkowski space as a fibrewise set over the real line, with the time coordinate as projection. Then the metric determines not a uniform structure but a fibrewise uniform structure, in the following sense.

As before, let B be a topological space and let X be a fibrewise set over B. By a *fibrewise uniform structure* on X I mean a filter Ω on $X \times X$ satisfying three conditions, as follows. The members of the filter are called *entourages* and are denoted by capital letters $D, E, F, \ldots$.

Condition (12.1). *Each entourage D contains the diagonal Δ of X.*

Condition (12.2). *If D is an entourage then for each point b of B there exists a neighbourhood W of b and an entourage E such that $X_W^2 \cap E \subset D^{-1}$.*

Condition (12.3). *If D is an entourage then for each point b of B there exists a neighbourhood W of b and an entourage E such that*

$$(X_W^2 \cap E) \circ (X_W^2 \cap E) \subset D.$$

It should be emphasized that E, in these two conditions, will generally depend on b as well as on D.

By a *fibrewise uniform space* we mean a fibrewise set, over a topological space, with a fibrewise uniform structure. Fibrewise uniform spaces over a point can be regarded as uniform spaces in the ordinary sense.

Clearly a uniform structure, in the ordinary sense, is also a fibrewise uniform structure. In particular the indiscrete uniform structure, in which the full set is the sole entourage, is also fibrewise uniform, hence the term *fibrewise indiscrete uniform structure*. We regard the base space B as fibrewise uniform over itself with this structure. More generally we regard the product $B \times T$ of B with a uniform space T as fibrewise uniform over B by taking the product of the indiscrete uniformity on B and the given uniformity on T.

If X is fibrewise uniform over B then $X_{B'}$ is fibrewise uniform over B' for each subset B' of B; the entourages of $X_{B'}$ are just the traces on $X_{B'}^2$ of the entourages of X. In particular the fibres X_b ($b \in B$) can be regarded as uniform spaces. In general, however, there will not be any neighbourhood W of b for which the fibrewise uniform structure on X_W is a uniform structure in the ordinary sense.

A fibrewise uniform structure on a fibrewise set X over B is said to be fibrewise discrete if for each point b of B there exists a neighbourhood W of b and an entourage D such that $(\xi, \eta) \in D$, where $\xi, \eta \in X_W$, implies $\xi = \eta$.

Let X be fibrewise uniform over B. Suppose that we have a fibrewise basis Λ for the entourage filter Ω, in the sense that for each entourage D and for each point b of B there exists a neighbourhood W of b and a member E of Λ for which $X_W^2 \cap E \subset D$. Then we describe Λ as a *fibrewise basis* for the fibrewise uniform structure, and refer to the members of Λ as the *fibrewise basic entourages*.

Reversing our point of view, let X be a fibrewise set over B. Suppose that we have a fibrewise basis, in the above sense, for a filter Ω on X^2. Let us refer to the members of Λ, without prejudice, as fibrewise basic entourages. Then Ω is a fibrewise uniform structure on X if and only if Λ satisfies the following three conditions.

Condition (12.4). *Each fibrewise basic entourage D contains the diagonal Δ.*

Condition (12.5). *If D is a fibrewise basic entourage then for each point b of B there exists a neighbourhood W of b and a fibrewise basic entourage E such that $X_W^2 \cap E \subset D^{-1}$.*

Condition (12.6). *If D is a fibrewise basic entourage then for each point* *b of B there exists a neighbourhood W of b and a fibrewise basic entourage E such that* $(X_W^2 \cap E) \circ (X_W^2 \cap E) \subset D$.

When these conditions are satisfied we refer to Ω as the fibrewise uniform structure *generated* by Λ. Although it is not strictly necessary to do so we shall, throughout the rest of this chapter, restrict ourselves to fibrewise uniform structures which admit a fibrewise basis of *symmetric* entourages: for these the second condition is superfluous.

Definition (12.7). *Let* $\phi: X \to Y$ *be a fibrewise function, where X and Y are fibrewise uniform over B. Then* ϕ *is* fibrewise uniformly continuous *if for each entourage E of Y and each point b of B there exists a neighbourhood W of b and an entourage D of X such that* $X_W^2 \cap D \subset \phi^{-2}E$.

It should be emphasized that fibrewise uniform continuity does not necessarily imply uniform continuity when X and Y are uniform spaces in the ordinary sense. When the fibrewise uniform structure on Y is generated by a fibrewise basis it is only necessary for the condition in (12.7) to be satisfied for the fibrewise basic entourages. Note that a fibrewise function is automatically fibrewise uniform continuous when the codomain has fibrewise indiscrete uniform structure.

Fibrewise uniform spaces over B form a category in which the morphisms are fibrewise uniformly continuous functions. The equivalences of the category are called *fibrewise uniform equivalences*. If X is fibrewise uniformly equivalent to $B \times T$ for some uniform space T we say that X is *trivial*, as a fibrewise uniform space. *Local triviality*, in the same sense, is defined similarly. Minkowski space provides an example of a trivial uniform space, the fibrewise basic entourages consisting of pairs of events which occur within ε of each other spatially, where $\varepsilon > 0$, regardless of time.

Induced fibrewise uniform structures are defined in the obvious way. Thus let Z be a fibrewise set over B. Let $\{X_r\}$ be a family of fibrewise uniform spaces over B, and let $\{\phi_r\}$ be a family of fibrewise functions, where $\phi_r: Z \to X_r$. The *induced fibrewise uniform structure* on Z is the coarsest for which all the functions ϕ_r are fibrewise uniformly continuous. The structure is characterized by the following property: for each fibrewise uniform space Y a fibrewise function $\theta: Y \to Z$ is fibrewise uniformly continuous if and only if each of the components $\phi_r\theta: Y \to X_r$ is fibrewise uniformly continuous.

In particular the *fibrewise uniform product structure* on the fibrewise product $\prod_B X_r$ is the structure induced by the family of projections

$$\pi_r: \prod_B X_r \to X.$$

Again, let $\phi: X' \to X$ be a fibrewise injection, where X and X' are fibrewise uniform over B. We say that ϕ is a *fibrewise uniform embedding*, if the structure of X' is induced from that of X by means of ϕ.

A particularly simple example is when X is fibrewise uniform over B and A is a subset of X. We regard A as fibrewise uniform over B with the structure induced by the inclusion, in which the entourages of A are just the traces on A^2 of the entourages of X.

Induced fibrewise uniform structures satisfy the usual transitivity conditions. Thus let

$$X \overset{\phi}{\to} Y \overset{\psi}{\to} Z$$

be fibrewise functions, where X and Y are fibrewise sets over B and where Z is a fibrewise uniform space over B. It makes no difference whether we give X fibrewise uniform structure by first taking the induced structure on Y, with respect to ψ, and then taking the induced structure on X, with respect to ϕ, or whether we give X the induced structure determined by the composition $\psi\phi$.

Definition (12.8). *The fibrewise uniform space X over B is fibrewise separated if the induced uniform structure on X_b is separated for each point b of B.*

For example the product $B \times T$ of B with the uniform space T is fibrewise separated whenever T is separated. Clearly each subspace of a fibrewise separated fibrewise uniform space is fibrewise separated. In fact we have

Proposition (12.9). *Let $\phi: X' \to X$ be a fibrewise uniform embedding, where X and X' are fibrewise uniform over B. If X is fibrewise separated then so is X'.*

Also we have

Proposition (12.10). *Let $\{X_r\}$ be a family of fibrewise separated fibrewise uniform spaces over B. Then the fibrewise uniform product $\prod_B X_r$ is also fibrewise separated.*

Proposition (12.11). *Let $\{B_j\}$ be an open covering of the topological space B. Then the fibrewise uniform space X over B is fibrewise separated if and only if X_{B_j} is fibrewise separated over B_j for each index j.*

13. Fibrewise uniform topology

We continue to work over a topological base space, and now show that each fibrewise uniformity determines a fibrewise topology on the same fibrewise set. Specifically, let X be fibrewise uniform over the topological space B. For each point b of B and each point x of X_b consider the family of subsets $D[x]$ of X, where D runs through the entourages of the fibrewise uniform structure. I refer to these, without prejudice, as the *fibrewise uniform neighbourhoods* of x. I assert that the system of tied filters generated by these families forms a fibrewise neighbourhood system on X and so defines a fibrewise topology, called the *fibrewise uniform topology*.

In fact the only condition which may not be entirely obvious is the coherence condition. To verify this, consider the fibrewise uniform neighbourhood $D[x]$ of x, where D is an entourage. By (12.3) there exists a neighbourhood W of b and a symmetric entourage E such that

$$(X_W^2 \cap E) \circ (X_W^2 \cap E) \subset D.$$

Then if $\xi \in X_W \cap E[x]$ we have $X_W \cap E[\xi] \subset D[x]$ and so $D[x]$ is a neighbourhood of ξ. Thus all the conditions for a fibrewise basis are satisfied.

For example, the fibrewise uniform topology determined by the fibrewise indiscrete uniformity is just the fibrewise indiscrete topology. Also for the fibrewise uniform space $B \times T$, where T is uniform, the fibrewise uniform topology is just the product topology of B with the uniform topology of T. If the fibrewise uniformity on X is given by a fibrewise basis for the entourage system then it is sufficient, in defining the fibrewise uniform topology, to use only fibrewise basic entourages.

A fibrewise topological space X over B is said to be *fibrewise uniformizable* if there exists a fibrewise uniform structure on X such that the underlying fibrewise uniform topology is the given fibrewise topology. For example, the fibrewise indiscrete topology is fibrewise uniformizable, as we have already noted. I do not know whether every fibrewise discrete topology is fibrewise uniformizable. However X is fibrewise uniformizable if X is an overlaying of B, in the sense of Fox (see §1).

To see this, let $\{W_j\}$ be a basis for the topology of B and let $\{\Gamma_j\}$ be a collection of families of subsets of X such that W_j is evenly covered by Γ_j and the overlaying condition is satisfied. With each family Γ_j we associate the subset D_j of $X_{W_j} \times X_{W_j}$ consisting of the union of the products $U \times U$ of the members U of Γ_j. Since the members of Γ_j are pairwise disjoint the condition $D_j \circ D_j = D_j$ is satisfied. It is easy to see that by taking as a fibrewise basis for the entourages the neighbourhoods $\bigcup D_j$ of the diagonal we obtain a fibrewise discrete uniform structure on X, and that the underlying fibrewise uniform topology is just the fibrewise discrete topology determined by the overlay structure.

Proposition (13.1). *Let $\phi: X \to Y$ be fibrewise uniformly continuous, where X and Y are fibrewise uniform over B. Then ϕ is continuous in the fibrewise uniform topology.*

To establish continuity at a given point x of a given fibre X_b, consider a fibrewise uniform neighbourhood $E[\phi(x)]$ of $\phi(x)$ in Y, where E is an entourage of Y. Now $X_W^2 \cap D \subset \phi^{-2}E$, for some neighbourhood W of b and entourage D of X, since ϕ is fibrewise uniformly continuous, and $X_W \cap D[x]$ is a neighbourhood of x in X. Since $X_W \cap D[x] \subset \phi^{-1}E[\phi(x)]$ this establishes continuity at the given point x and thereby proves (13.1).

Corollary (13.2). *Let $\phi: X \to Y$ be a fibrewise uniform equivalence, where X and Y are fibrewise uniform over B. Then ϕ is a fibrewise topological equivalence, in the fibrewise uniform topology.*

It follows, in particular, that if X is trivial as a fibrewise uniform space over B then X is trivial as a fibrewise topological space over B.

Our main result in the opposite direction to (13.2) is given in §17, but the following special case may be proved at this stage.

Proposition (13.3). *Let $s: B \to X$ be a (continuous) section, where X is fibrewise uniform over B. Then s is fibrewise uniformly continuous.*

For let D be an entourage of X and let b be a point of B. Then $(X_W^2 \cap E) \circ (X_W^2 \cap E) \subset D$ for some neighbourhood W of b and symmetric entourage E. Since $X_W \cap E[x]$ is a neighbourhood of $x = s(b)$, there exists a neighbourhood $W' \subset W$ of b such that

$$W' \subset s^{-1}(X_W \cap E[x]) = W \cap s^{-1}E[x].$$

Then

$$W' \times W' \subset s^{-2}(E[x] \times E[x]) \subset s^{-2}D,$$

and so s is fibrewise uniformly continuous, as asserted. It follows at once that fibrewise constant functions between fibrewise uniform spaces are fibrewise uniformly continuous.

Let Z be a fibrewise set over B. Let $\{X_r\}$ be a family of fibrewise uniform spaces over B, and let $\{\phi_r\}$ be a family of fibrewise functions, where $\phi_r: Z \to X_r$. It is easy to check that the fibrewise uniform topology on Z

determined by the induced fibrewise uniform structure coincides with the fibrewise topology induced by the same functions with respect to the underlying fibrewise topological structures.

In particular, the fibrewise uniform topology determined by the fibrewise uniform product of fibrewise uniform spaces coincides with the fibrewise topological product of the underlying fibrewise topological spaces.

Again, if X is fibrewise uniform over B and A is a subset of X then the fibrewise uniform topology on A, as a fibrewise set over B, is the same whether we first take the induced fibrewise uniform structure and then the fibrewise uniform topology or first take the fibrewise uniform topology and then the induced fibrewise topology.

We turn now to a series of results which help to give a better insight into the relation between the fibrewise uniform structure and the fibrewise uniform topology.

Proposition (13.4). *The fibrewise uniform space X over B is fibrewise separated if and only if the underlying fibrewise topological space is fibrewise Hausdorff.*

If X is fibrewise Hausdorff each fibre is Hausdorff, and therefore separated, in the induced uniform structure, and so X is fibrewise separated in the fibrewise uniform structure. Conversely suppose that X is fibrewise separated. Let b be a point of B and let x, x' be distinct points of X_b. Then there exists an entourage D to which (x, x') does not belong. Then $X_W \cap E[x]$, $X_W \cap E[x']$ are disjoint neighbourhoods of x, x', where W is a neighbourhood of b and E is a symmetric entourage such that

$$(X_W^2 \cap E) \circ (X_W^2 \cap E) \subset D,$$

and so X is fibrewise Hausdorff.

Proposition (13.5). *Let X be fibrewise uniform over B. Then for each point b of B and each entourage D there exists a neighbourhood W of b and an entourage E such that*

$$X_W^2 \cap E \subset \operatorname{Int} D, \qquad X_W^2 \cap \operatorname{Cl} E \subset D.$$

By the axioms there exists a neighbourhood W of b and a symmetric entourage E such that $(X_W^2 \cap E) \circ (X_W^2 \cap E) \circ (X_W^2 \cap E) \subset D$. If $(x, x') \in E$, where $x, x' \in X_W$, then

$$(X_W \cap E[x]) \times (X_W \cap E[x'])$$

is a neighbourhood of (x, x') contained in D; this proves the first assertion.

Again if $(x, x') \in \mathrm{Cl}\, E$, where $x, x' \in X_W$, then the neighbourhood

$$(X_W \cap E[x]) \times (X_W \cap E[x'])$$

of (x, x') meets E, and so $(x, x') \in (X_W^2 \cap E) \circ (X_W^2 \cap E) \circ (X_W^2 \cap E) \subset D$. Thus $X_W^2 \cap \mathrm{Cl}\, E \subset D$ as asserted.

It follows at once from (13.5) that each entourage is a neighbourhood of the diagonal Δ. Also we have

Proposition (13.6). *Let X be fibrewise uniform over B. Then X is fibrewise regular, in the fibrewise uniform topology.*

For let x be a point of the fibre X_b. For each entourage D there exists, by (13.5), a neighbourhood W of b and a closed entourage D' such that $X_W^2 \cap D' \subset D$. Then $X_W \cap D'[x]$ is a closed neighbourhood of x in X_W such that $X_W \cap D'[x] \subset D[x]$, as required.

14. The Cauchy condition

Recall that a subset M of a set X is said to be *D-small*, where $D \subset X^2$, if M^2 is contained in D. When $D = A^2$, where $A \subset X$, this simply means $M \subset A$. Observe that if M_i is D-small $(i = 1, 2)$ then $M_1 \cup M_2$ is $D \circ D$-small provided M_1 meets M_2. We continue to work over a topological space B.

Definition (14.1). *Let X be fibrewise uniform over B. A b-filter $\mathscr{F}$ on X, where $b \in B$, is* Cauchy *if $\mathscr{F}$ contains a D-small member for each entourage D.*

For example, every b-filter is necessarily Cauchy when X has fibrewise indiscrete uniform structure. Note that it is sufficient for the condition in (14.1) to be satisfied for the fibrewise basic entourages of a fibrewise basis.

Proposition (14.2). *Let X be fibrewise uniform over B. Let $\mathscr{F}$ be a convergent b-filter on X $(b \in B)$. Then $\mathscr{F}$ is a Cauchy b-filter.*

For let D be an entourage of X. Then $(X_W^2 \cap E) \circ (X_W^2 \cap E) \subset D$ for some neighbourhood W of b and symmetric entourage E of X. If the b-filter $\mathscr{F}$ on X converges to some point x of X then the neighbourhood $X_W \cap E[x]$ of x is a member of $\mathscr{F}$. But $X_W \cap E[x]$ is $(X_W^2 \cap E) \circ (X_W^2 \cap E)$-small and so D-small as required.

Proposition (14.3). *Let $\phi: X \to Y$ be fibrewise uniformly continuous, where X and Y are fibrewise uniform over B. If $\mathcal{F}$ is a Cauchy b-filter on X where $b \in B$ then $\phi_* \mathcal{F}$ is a Cauchy b-filter on Y.*

For let E be an entourage of Y. Then $X_W^2 \cap D \subset \phi^{-2} E$ for some neighbourhood W of b and entourage D of X. If $M \in \mathcal{F}$ is E-small then $\phi(X_W \cap M) = Y_W \cap \phi M \in \phi_* \mathcal{F}$ is $\phi^2(X_W^2 \cap D)$-small and so E-small as required.

Proposition (14.4). *Let $\phi: X \to Y$ be a fibrewise function, where X is a fibrewise set and Y a fibrewise uniform space over B. Let $\mathcal{G}$ be a b-filter on Y ($b \in B$) such that $\phi^* \mathcal{G}$ is defined as a b-filter on X. If $\mathcal{G}$ is Cauchy on Y then $\phi^* \mathcal{G}$ is Cauchy on X, in the induced fibrewise uniform structure.*

For let D be an entourage of X. Then $X_W^2 \cap \phi^{-2} E \subset D$ for some neighbourhood W of b and entourage E of Y. If M is an E-small member of $\mathcal{G}$ then $X_W \cap \phi^{-1} M$ is a D-small member of $\phi^* \mathcal{G}$. In particular we have

Corollary (14.5). *Let X be fibrewise uniform over B and let A be a subspace of X. Let $\mathcal{F}$ be a b-filter on X ($b \in B$) such that the trace $\mathcal{F}_A$ of $\mathcal{F}$ on A is defined as a b-filter. If $\mathcal{F}$ is Cauchy then $\mathcal{F}_A$ is also Cauchy.*

For a given fibrewise uniform space over B the minimal elements of the class of Cauchy b-filters ($b \in B$), with respect to the relation of refinement, are called *minimal Cauchy b-filters*. We have

Proposition (14.6). *Let X be fibrewise uniform over B. Let $\mathcal{F}$ be a Cauchy b-filter on X ($b \in B$). Then there exists one and only one minimal Cauchy b-filter $\mathcal{F}'$ on X such that $\mathcal{F}$ is a refinement of $\mathcal{F}'$.*

To see this consider the b-filter $\mathcal{F}'$ on X generated by the subsets $D[M]$, where M runs through the members of $\mathcal{F}$ and D runs through the entourages of X. I assert that $\mathcal{F}'$ is Cauchy. For let D be an entourage of X. Then $(X_W^2 \cap E) \circ (X_W^2 \cap E) \circ (X_W^2 \cap E) \subset D$ for some neighbourhood W of b and symmetric entourage E of X. Since $\mathcal{F}$ is Cauchy there exists an E-small member M of $\mathcal{F}$. Then $X_W \cap E[M]$ is

$$(X_W^2 \cap E) \circ (X_W^2 \cap E) \circ (X_W^2 \cap E)\text{-small}$$

and so D-small. Thus $\mathcal{F}'$ is Cauchy. Also $\mathcal{F}$ is a refinement of $\mathcal{F}'$.

I now assert that $\mathcal{F}'$ is minimal. For let $\mathcal{G}$ be a Cauchy b-filter refined by $\mathcal{F}'$. Let M be a member of $\mathcal{F}'$ and let D be an entourage. Since $\mathcal{G}$ is

Cauchy there exists a D-small member N, say, of $\mathscr{G}$. Now N meets M, since both sets belong to $\mathscr{F}'$. Therefore $D[M] \in \mathscr{G}$, since $N \subset D[M]$, and so $\mathscr{G} = \mathscr{F}'$. This shows that $\mathscr{F}'$ is minimal and also establishes uniqueness.

Thus to each Cauchy b-filter $\mathscr{F}$ there corresponds the unique Cauchy b-filter $\mathscr{F}'$ satisfying the minimality condition. If the filter is generated by a fibrewise basis, or if the fibrewise uniform structure of X is generated by a fibrewise basis, then a fibrewise basis for $\mathscr{F}'$ consists of the subsets $D[M]$ where M runs through the fibrewise basic members of $\mathscr{F}$ and D runs through the fibrewise basic entourages of X.

In particular, take $\mathscr{F}$ to be the principal filter $\mathscr{E}_x$ generated by a given point x of X_b. Then we obtain

Proposition (14.7). *Let X be fibrewise uniform over B. For each point b of B and each point x of X_b the neighbourhood filter $\mathscr{N}_x$ of x in X is a minimal Cauchy b-filter.*

Proposition (14.8). *Let $\mathscr{F}$ be a Cauchy b-filter $(b \in B)$ on the fibrewise uniform space X over B. If x is an adherence point of $\mathscr{F}$ then x is a limit point of $\mathscr{F}$.*

For suppose that x is an adherence point of $\mathscr{F}$. Let $\mathscr{G}$ be a common refinement of the Cauchy b-filters $\mathscr{F}$ and $\mathscr{N}_x$. If $\mathscr{F}'$ is the unique minimal Cauchy b-filter refined by $\mathscr{F}$ then $\mathscr{F}' = \mathscr{N}_x$, since $\mathscr{F}'$ and $\mathscr{N}_x$ are both minimal Cauchy b-filters refined by $\mathscr{G}$. Thus $\mathscr{F}$ converges to x, as asserted. We at once deduce

Corollary (14.9). *Let $\mathscr{F}$ be a Cauchy b-filter on X and let $\mathscr{G}$ be a refinement of $\mathscr{F}$. Then any limit point of $\mathscr{G}$ is also a limit point of $\mathscr{F}$.*

Definition (14.10). *The fibrewise uniform space X over B is fibrewise complete if each Cauchy tied filter on X is convergent.*

For example, fibrewise indiscrete uniform spaces and fibrewise uniform spaces derived from overlayings are fibrewise complete. Fibrewise uniform spaces with fibrewise compact fibrewise uniform topology are also fibrewise complete, since every tied filter has an adherence point and so every Cauchy tied filter has a limit point.

Proposition (14.11). *Let $\phi: X \rightarrow Y$ be a fibrewise surjection, where X is a fibrewise set and Y is a fibrewise uniform space over B. Suppose that X has the fibrewise uniform structure induced by ϕ. Then Y is fibrewise complete whenever X is fibrewise complete.*

This follows from (14.3) and (14.4).

Proposition (14.12). *Let $\phi: X \rightarrow X'$ be a closed fibrewise uniform embedding, where X and X' are fibrewise uniform over B. If X' is fibrewise complete then so is X.*

For let $\mathscr{F}$ be a Cauchy b-filter on X, where $b \in B$. Then $\phi_* \mathscr{F}$ is a Cauchy b-filter on X', by (14.3). Suppose that X' is fibrewise complete. Then $\phi_* \mathscr{F}$ converges to a point x' of X'_b. Now ϕX is closed in X', since ϕ is closed, and $\phi X \in \phi_* \mathscr{F}$, since $X \in \mathscr{F}$. Hence $x' \in \phi X$. If $\phi(x) = x'$, where $x \in X_b$, then $\mathscr{F}$ converges to x, since ϕ is injective. Therefore X is fibrewise complete, as asserted.

It follows, in particular, that closed subspaces of fibrewise complete fibrewise uniform spaces are also fibrewise complete. In the other direction we have

Proposition (14.13). *Let $\phi: X \rightarrow Y$ be a fibrewise uniform embedding, where X and Y are fibrewise uniform over B. If X is fibrewise complete and Y is fibrewise separated then ϕ is closed.*

For let y be an adherence point of ϕX, where $y \in Y_b$. The neighbourhood filter $\mathscr{N}_y$ of y is a Cauchy b-filter, by (14.7). Since y is an adherence point of ϕX the trace of $\mathscr{N}_y$ on ϕX is defined and so $\phi^* \mathscr{N}_y$ is defined, as a Cauchy b-filter on X. If X is fibrewise complete then $\phi^* \mathscr{N}_y$ converges to a point $x \in X_b$. Then $\phi(x)$ and y are both adherence points of $\mathscr{N}_y$. If, further, Y is fibrewise separated then $\phi(x) = y$, by (3.3) and (13.4), so that ϕX is closed. This proves (14.13).

It follows, in particular, that fibrewise complete subspaces of fibrewise separated fibrewise uniform spaces are closed, in the fibrewise uniform topology.

Proposition (14.14). *Let X be fibrewise uniform over B and let A be a dense subspace of X. Suppose that each Cauchy tied filter on A is convergent as a tied filter on X. Then X is fibrewise complete.*

It is sufficient to show that each minimal Cauchy tied filter $\mathscr{F}$ on X is convergent. Since A is dense in X and since each member M of $\mathscr{F}$ has a non-empty interior, by (13.5), the trace $\mathscr{F}_A$ of $\mathscr{F}$ on A is a Cauchy tied filter on A and so converges to some point x of X. But $u_*\mathscr{F}_A$ is a refinement of $\mathscr{F}$, where $u\colon A \subset X$, and so x adheres to $\mathscr{F}$. Hence $\mathscr{F}$ converges to x, since $\mathscr{F}$ is Cauchy. This proves (14.14).

Proposition (14.15). *Let X be fibrewise uniform over B and let A be a dense subspace of X. Let $\phi\colon A \to Y$ be fibrewise uniformly continuous, where Y is fibrewise separated and fibrewise complete over B. Then there exists one and only one fibrewise uniformly continuous function $\psi\colon X \to Y$ such that $\psi\,|\,A = \phi$.*

First observe that Y, as a fibrewise topological space, is both fibrewise regular, by (13.6), and fibrewise Hausdorff, by (13.4). To define ψ at a given point x of X_b, where $b \in B$, we apply ϕ_* to the trace on A of the neighbourhood filter $\mathscr{N}_x$ of x in X. Since $\mathscr{N}_x$ is a Cauchy b-filter so is its trace and hence so is the image of the trace under ϕ_*, by (14.3). Since Y is fibrewise complete the image converges to a point of Y, which is unique since Y is fibrewise Hausdorff, and we take this to be the value of ψ at x. The continuity of ψ, thus defined, follows at once from (4.5).

To show that ψ is fibrewise uniformly continuous, let E be an entourage of Y, and let $b \in B$. By (13.5) $Y^2_{W''} \cap E' \subset E$ for some neighbourhood W'' of b and some closed entourage E' of Y. Since $\phi = \psi\,|\,A$ is fibrewise uniformly continuous there exists a neighbourhood $W' \subset W''$ of b and an entourage D' of X such that $A^2_{W'} \cap D' \subset \phi^{-2}E'$. By (13.5) there exists a neighbourhood $W \subset W'$ of b and an entourage D of X such that $X^2_W \cap \bar{D} \subset D'$. Now $X^2_W \cap D \subset \psi^{-2}E'$ and so $X^2_W \cap \bar{D} \subset \psi^{-2}E'$, since ψ^2 is continuous and E' is closed. Consequently $X^2_W \cap \bar{D} \subset \psi^{-2}E$; since $\bar{D}$ is an entourage of X this shows that ψ is fibrewise uniformly continuous, as asserted.

Corollary (14.16). *Let X_1, X_2 be fibrewise separated and fibrewise complete over B. Let A_1, A_2 be dense subspaces of X_1, X_2, respectively. Then any fibrewise uniform equivalence ϕ between A_1 and A_2 can be extended to a fibrewise uniform equivalence ψ between X_1 and X_2.*

For the function $A_1 \to X_2$ determined by ϕ is fibrewise uniformly continuous and so extends to a fibrewise uniformly continuous function $\psi \colon X_1 \to X_2$. Similarly the function $A_2 \to X_1$ determined by ϕ^{-1} extends to a fibrewise uniformly continuous function $\psi' \colon X_2 \to X_1$. By the uniqueness of the fibrewise extension, $\psi' \circ \psi$ is the identity on X_2. Since both functions are fibrewise uniformly continuous we conclude that ψ is a fibrewise uniform equivalence with inverse ψ'.

15. Fibrewise completion

We continue to work over a topological space B. The main purpose of this section is to prove

Proposition (15.1). *Let X be fibrewise uniform over B. Then there exists a fibrewise separated and fibrewise complete fibrewise uniform space $\hat{X}_B$, and a fibrewise uniformly continuous function $\rho \colon X \to \hat{X}_B$, satisfying the following condition.*

Condition (15.2). *For each fibrewise separated and fibrewise complete fibrewise uniform space Y over B and each fibrewise uniformly continuous function $\phi \colon X \to Y$ there exists one and only one fibrewise uniformly continuous function $\psi \colon \hat{X}_B \to Y$ such that $\psi \circ \rho = \phi$.*

The condition is in the nature of a universal property and so, for formal reasons, we have uniqueness, up to an appropriate notion of equivalence. Specifically if $\rho' \colon X \to \hat{X}'_B$ also satisfies (15.2) then there exists a fibrewise uniform equivalence $\theta \colon \hat{X}_B \to \hat{X}'_B$ such that $\theta \circ \rho = \rho'$. Consequently we shall refer to $\hat{X}_B$ as the *fibrewise separated completion* of X and to ρ as the *canonical fibrewise function*.

The direct image ρX of X in $\hat{X}_B$ is also of considerable interest. We shall denote it by $\tilde{X}_B$ and denote by $\alpha \colon X \to \tilde{X}_B$ the canonical function given by ρ. Clearly $\tilde{X}_B$ is fibrewise separated, since $\hat{X}_B$ is fibrewise separated. For reasons which will become clearer as we proceed we shall refer to $\tilde{X}_B$ as the *maximal fibrewise separated quotient* of X.

In addition to the universal property (15.2) we shall show that $\tilde{X}_B$ is dense in $\hat{X}_B$, and that X has the fibrewise uniform structure induced from $\hat{X}_B$ by means of ρ. Consequently X also has the fibrewise uniform structure induced from $\tilde{X}_B$ by means of α. In fact, as we shall see, the underlying fibrewise topological space of $\tilde{X}_B$ is fibrewise topologically equivalent to the maximal fibrewise Hausdorff quotient of X, in the sense of §5.

When X is fibrewise separated it follows that ρ is injective and so a fibrewise uniform embedding. In this case, therefore, we may identify $\tilde{X}_B$ with X and regard X as a subspace of $\hat{X}_B$ with the canonical function as the inclusion. Similarly when X is fibrewise complete the inclusion $\tilde{X}_B \to \hat{X}_B$ is a fibrewise uniform equivalence, from the universal property and (14.11), and so we may identify $\hat{X}_B$ with $\tilde{X}_B$ and ρ with α.

In the next section we shall show that both fibrewise separated completion and maximal fibrewise separated quotient are functorial, with respect to fibrewise uniformly continuous functions.

The proof of (15.1) is in several steps. First of all we define $\hat{X}_B$ as a fibrewise set so that the fibre over a given point b of B consists of the minimal Cauchy b-filters $\mathscr{F}$ on X, and we define the fibrewise uniform structure of $\hat{X}_B$ as follows. For each symmetric entourage D of X let D^* denote the subset of $\hat{X}_B^2$ consisting of pairs $((b_1, \mathscr{F}_1), (b_2, \mathscr{F}_2))$ of minimal Cauchy tied filters on X such that $\mathscr{F}_1$ and $\mathscr{F}_2$ have a common D-small member. We shall show that a fibrewise uniform structure on $\hat{X}_B$ is defined by taking the family of entourages to be that generated by the family of subsets D^*, where D runs through the symmetric entourages of X. Thus we have to show that the conditions (12.4)–(12.6) for a fibrewise basis are satisfied.

First observe that if $(b, \mathscr{F})$ is a minimal Cauchy b-filter where $b \in B$ then the pair $((b, \mathscr{F}), (b, \mathscr{F}))$ belongs to D^*, for each symmetric entourage D, since $\mathscr{F}$ contains a D-small member by the Cauchy condition. Since the symmetric entourages form a fibrewise basis this shows that each entourage of $\hat{X}_B$ contains the diagonal. Thus (12.4) is satisfied.

Of the remaining conditions (12.5) follows directly from the corresponding property in the case of X. As for (12.6), let D be a symmetric entourage of X and let b be a point of B. Then $(X_W^2 \cap E) \circ (X_W^2 \cap E) \subset D$, for some neighbourhood W of b and some symmetric entourage E of X. I assert that $((\hat{X}_B)_W^2 \cap E^*) \circ ((\hat{X}_B)_W^2 \cap E^*) \subset D^*$. For let $(b_i, \mathscr{F}_i)$ be a minimal Cauchy b_i-filter on X $(i = 1, 2, 3)$, where $b_i \in W$. Suppose that both pairs

$$((b_1, \mathscr{F}_1), (b_2, \mathscr{F}_2)), \qquad ((b_2, \mathscr{F}_2), (b_3, \mathscr{F}_3))$$

are contained in E^*. Then there exist E-small subsets M, N of X such that $M \in \mathscr{F}_1 \cap \mathscr{F}_2$, $N \in \mathscr{F}_2 \cap \mathscr{F}_3$. Now M meets N, since both sets are members of $\mathscr{F}_2$. Therefore $X_W \cap (M \cup N)$ is $(X_W^2 \cap E) \circ (X_W^2 \cap E)$-small and so D-small. But $X_W \cap (M \cup N)$ is common to both $\mathscr{F}_1$ and $\mathscr{F}_3$, and so the pair

$$((b_1, \mathscr{F}_1), (b_3, \mathscr{F}_3))$$

is contained in D^*. This proves the assertion and shows that (12.6) is satisfied.

Next let us show that $\hat{X}_B$ is fibrewise separated. Consider the minimal Cauchy b-filter $\mathscr{F}_i$ $(i = 1, 2)$ $(b \in B)$. Suppose that the pair

$$((b, \mathscr{F}_1), (b, \mathscr{F}_2))$$

is contained in D^* for each symmetric entourage D of X. The unions $M_1 \cup M_2$, where $M_1 \in \mathscr{F}_1$ and $M_2 \in \mathscr{F}_2$, form the fibrewise basis of a b-filter $\mathscr{F}_0$ on X which is refined by both $\mathscr{F}_1$ and $\mathscr{F}_2$. Now $\mathscr{F}_0$ is Cauchy since for each symmetric entourage D of X there exists a common D-small member of $\mathscr{F}_1$ and $\mathscr{F}_2$. But then $\mathscr{F}_0$ coincides with both $\mathscr{F}_1$ and $\mathscr{F}_2$, by minimality, and so $\mathscr{F}_1 = \mathscr{F}_2$. Thus $\hat{X}_B$ is fibrewise separated, as asserted.

The canonical function $\rho: X \to \hat{X}_B$ is defined so as to associate with each point x of X_b $(b \in B)$ the minimal Cauchy b-filter $\mathscr{N}_x$. Clearly ρ is fibrewise. I assert that X has the induced fibrewise uniform structure. To see this first observe that

(15.3) $$\rho^{-2}D^* \subset D$$

for each symmetric entourage D of X. For if $(\rho(\xi), \rho(\eta)) \in D^*$, where $\xi, \eta \in X$, there exists a D-small common neighbourhood of ξ and η so that $(\xi, \eta) \in D$, as required. I also assert that

(15.4) $$D \subset \rho^{-2}(D \circ D \circ D)^*.$$

For if $(\xi, \eta) \in D$, where $\xi, \eta \in X$, then $D[\xi] \cap D[\eta]$ constitutes a common neighbourhood of ξ and η. But $D[\xi] \cap D[\eta]$ is $D \circ D \circ D$-small and so $(\rho(\xi), \rho(\eta)) \in (D \circ D \circ D)^*$, as required. It follows at once from (15.3) and (15.4) that X has the induced fibrewise uniform structure.

Next we show, at one and the same time, that $\rho X = \tilde{X}_B$ is dense in $\hat{X}_B$ and that $\hat{X}_B$ is fibrewise complete. For this purpose, let $\mathscr{F}$ be a minimal Cauchy b-filter $(b \in B)$ on X and consider the fibrewise uniform neighbourhood $D^*[\mathscr{F}]$ of $\mathscr{F}$, where D is a symmetric entourage of X. The trace of this neighbourhood on $\tilde{X}_B$ is the set of points $\rho(x)$ $(x \in X)$ such that $(\rho(x), (b, \mathscr{F}))$ is contained in D^*. This condition signifies that there exists a D-small neighbourhood of x belonging to $\mathscr{F}$, or, equivalently, that x belongs to the interior of some D-small member of $\mathscr{F}$.

So let $M \subset X$ be the union of the interiors of all D-small members of $\mathscr{F}$. Then $M \in \mathscr{F}$ and so $\tilde{X}_B \cap D^*[\mathscr{F}] = \rho M$, by what we have just shown. This implies that $\tilde{X}_B \cap D^*[\mathscr{F}]$ is non-empty, hence $\tilde{X}_B$ is dense in $\hat{X}_B$. It also implies that the trace of $D^*[\mathscr{F}]$ on $\tilde{X}_B$ belongs to the minimal Cauchy b-filter $\rho_*\mathscr{F}$ on $\tilde{X}_B$, and so $\rho_*\mathscr{F}$ converges in $\hat{X}_B$ to the point $\mathscr{F}$.

Now let $\mathscr{F}$ be any Cauchy b-filter on $\hat{X}_B$. Then $\rho^*\mathscr{F}$ is a Cauchy b-filter on X, by (14.4), since the fibrewise uniform structure of X is induced from that of $\hat{X}_B$. Let $\mathscr{G}$ be the minimal Cauchy b-filter on X refined by $\rho^*\mathscr{F}$.

Then $\rho_* \mathscr{G}$ is a Cauchy b-filter on $\hat{X}_B$, by (14.3), and is refined by $\rho_* \rho^* \mathscr{F} = \mathscr{F}$. Since $\rho_* \mathscr{G}$ converges in $\hat{X}_B$, as we have just seen, so does $\mathscr{F}$ itself, by (14.9). This establishes that $\hat{X}_B$ is fibrewise complete.

Finally we have to show that the universal property (15.2) is satisfied. So let Y be fibrewise separated and fibrewise complete over B and let $\phi \colon X \to Y$ be fibrewise uniformly continuous. We begin by showing that there exists a unique fibrewise uniformly continuous function $\psi_0 \colon \tilde{X}_B \to Y$ such that $\phi = \psi_0 \circ \alpha$. In fact, since ϕ is continuous we have at once that $\phi(x) = \lim \phi_* \mathscr{N}_x$, for each point x of X_b $(b \in B)$, and so if we write

$$\psi_0(\alpha(x)) = \lim \phi_* \mathscr{N}_x$$

we obtain a fibrewise function ψ_0 such that $\psi_0 \circ \alpha = \phi$. Moreover ψ_0 is fibrewise uniformly continuous. For let E be an entourage of Y and let $b \in B$. Then $X_W^2 \cap D \subset \phi^{-2}E$, for some neighbourhood W of b and symmetric entourage D of X, and then (15.3) implies that $(\hat{X}_B)_W^2 \cap D^* \subset (\psi_0 \alpha)^{-2}E$. Finally we fibrewise extend ψ_0 from $\tilde{X}_B$ to $\hat{X}_B$, as in (14.15), so as to obtain the required fibrewise uniformly continuous function ψ such that $\psi \circ \rho = \phi$. Since ψ_0 is uniquely defined, and since $\tilde{X}_B$ is dense in $\hat{X}_B$, the uniqueness of ψ follows from (2.6). This completes the proof of (15.1).

16. Functoriality

The fibrewise separated completion is functorial in the following sense.

Proposition (16.1). *Let $\phi \colon X \to X'$ be fibrewise uniformly continuous, where X and X' are fibrewise uniform over B. Then there exists one and only one fibrewise uniformly function $\hat{\phi}$ such that $\hat{\phi} \circ \rho = \rho' \circ \phi$, as shown below*

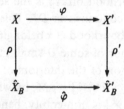

Here ρ and ρ' are canonical. The existence and uniqueness of $\hat{\phi}$ follow by applying (15.2) to $\rho'\phi$. Specifically, $\hat{\phi}$ transforms each Cauchy b-filter $\mathscr{F}$ on X into the Cauchy b-filter $\phi_* \mathscr{F}$ on X'. We refer to ϕ as the *fibrewise separated completion* of ϕ. Clearly the fibrewise separated completion of a composition is the composition of the fibrewise separated completions and hence the fibrewise separated completion of a fibrewise uniform equivalence is again a fibrewise uniform equivalence.

We now turn our attention to maximal fibrewise separated quotients and prove

Proposition (16.2). *Let X be fibrewise uniform over B. Then for each fibrewise separated fibrewise uniform space Y and each fibrewise uniformly continuous function $\phi: X \to Y$ there exists one and only one fibrewise uniformly continuous function $\psi: \tilde{X}_B \to Y$ such that $\psi \circ \alpha = \phi$.*

For by composing ϕ with the canonical $\sigma: Y \to \hat{Y}_B$ we obtain a fibrewise uniformly continuous function $\phi': X \to \hat{Y}_B$. By the universal property there exists a fibrewise uniformly continuous function $\psi': \hat{X}_B \to \hat{Y}_B$ such that $\psi' \circ \rho = \phi' = \sigma \circ \phi$. But σ is a fibrewise uniform embedding, since Y is fibrewise separated, and so the restriction of ψ' to $\tilde{X}_B$ determines a fibrewise uniformly continuous function $\psi: \tilde{X}_B \to \tilde{Y}_B$, as required. The uniqueness is obvious. We go on to prove

Proposition (16.3). *Let $\phi: X \to Y$ be a fibrewise surjection, where X is a fibrewise set and Y is a fibrewise separated fibrewise uniform space over B. If X has the fibrewise uniform structure induced from that of Y by means of ϕ then the corresponding function $\psi: \tilde{X}_B \to Y = \tilde{Y}_B$ is a fibrewise uniform equivalence.*

Here ψ is related to ϕ as in (16.2). Obviously ψ is surjective, since ϕ is surjective. I assert that ψ is injective, and so is bijective. For suppose that $\phi(\xi) = \phi(\eta)$, where $\xi, \eta \in X_b$ ($b \in B$). Then the pair (ξ, η) is contained in every entourage of X, since the pair $(\phi(\xi), \phi(\eta))$ is contained in ΔY, and so $\alpha(\xi) = \alpha(\eta)$ as required. Moreover the entourages of Y are the images under ϕ^2 of the entourages of X and so are the images under ψ^2 of the entourages of $\tilde{X}_B$. Thus both ψ and ψ^{-1} are fibrewise uniformly continuous, i.e. ψ is a fibrewise uniform equivalence, as asserted.

Proposition (16.4). *Let Z be a fibrewise set over B. Let $\{X_r\}$ be a family of fibrewise uniform spaces over B and let $\{\phi_r\}$ be a family of fibrewise functions, where $\phi_r: Z \to X_r$. Suppose that Z has the induced fibrewise uniform structure with respect to $\{\phi_r\}$. Then the fibrewise separated completion $\hat{Z}_B$ has the induced fibrewise uniform structure with respect to $\{\hat{\phi}_r\}$, where $\hat{\phi}_r: \hat{Z}_B \to (X_r)\hat{_B}$. Moreover if $\theta_r = \rho_r \circ \phi_r$, where $\rho_r: X_r \to (X_r)\hat{_B}$ is canonical, then $\hat{Z}_B$ can be identified with the closure in the fibrewise* uniform product $\prod_B (X_r)\hat{_B}$ of the image of Z under the function with rth component θ_r.

For consider the commutative diagram shown below.

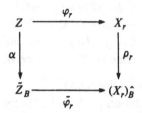

I assert that the fibrewise uniform structure which $\tilde{Z}_B$ obtains as a subspace of $\hat{Z}_B$ coincides with the induced structure which $\tilde{Z}_B$ obtains from $\{\tilde{\phi}_r\}$. In view of (16.3) it is sufficient to show that the induced structure is fibrewise separated. Now by transitivity of induced structures, the structure on Z induced by the induced structure on $\tilde{Z}_B$ coincides with the structure on Z induced by $\{\rho_r \circ \phi_r\}$. So if ξ, η are points of Z_b ($b \in B$) such that $\rho_r \phi_r(\xi) = \rho_r \phi_r(\eta)$ for each index r then the pair (ξ, η) belongs to each entourage of the induced structure on Z and so $\alpha(\xi) = \alpha(\eta)$, as required.

Using the $\{\tilde{\phi}_r\}$ therefore, we can embed $\tilde{Z}_B$ as a subspace of the fibrewise uniform product $\prod_B (X_r)\tilde{_B}$. Now the latter is a dense subspace of the fibrewise uniform product $\prod_B (X_r)\hat{_B}$, since the maximal fibrewise separated quotient is a dense subspace of the fibrewise separated completion. Regarding $\tilde{Z}_B$ as a subspace of $\prod_B (X_r)\hat{_B}$, therefore, we see that the closure is fibrewise separated and fibrewise complete, from (14.12). By uniqueness we can identify the closure with $\hat{Z}_B$ in the way described.

Corollary (16.5). *Let A be a subspace of the fibrewise uniform space X over B. Then $\hat{u}: \hat{A}_B \to \hat{X}_B$, where $u: A \subset X$, determines a fibrewise uniform equivalence between $\hat{A}_B$ and the closure of $\tilde{A}_B \subset \tilde{X}_B$ in $\hat{X}_B$.*

Corollary (16.6). *Let $\{X_r\}$ be a family of fibrewise uniform spaces over B. Then the fibrewise separated completion of the fibrewise uniform product $\prod_B X_r$ is canonically fibrewise uniformly equivalent to the fibrewise uniform product $\prod_B (X_r)\hat{_B}$ of the family of fibrewise separated completions $(X_r)\hat{_B}$.*

17. Fibrewise compactness and precompactness

Let X be fibrewise topological over B, and consider the neighbourhood filter of the diagonal Δ in X^2. One of the main results of the theory of uniform spaces is that the filter constitutes a uniform structure on X whenever X is compact regular. We generalize this by proving†

Proposition (17.1). *Let X be a fibrewise compact and fibrewise regular space over B, with B regular. Then there exists a unique fibrewise uniform structure on X, compatible with the fibrewise topology, in which the entourages are the neighbourhoods of the diagonal.*

First we show that in any fibrewise uniform structure on X, compatible with the fibrewise topology, there exists for each point b of B and each neighbourhood N of the diagonal Δ a neighbourhood W of b and an entourage D such that $X_W^2 \cap D \subset N$.

For suppose, to obtain a contradiction, that there exists a point b of B and a neighbourhood N of Δ, such that $X_W^2 \cap D \subset N$ is false for every neighbourhood W of b and every entourage D. Without real loss of generality we may suppose N to be open. Then the complement CN is closed and the intersections $X_W^2 \cap D \cap CN$ are non-empty, where W runs through the neighbourhoods of b and D runs through the entourages of the structure. The intersections therefore generate a (b, b)-filter $\mathscr{F}$ on CN. Also X^2 is fibrewise compact over B^2, since X is fibrewise compact over B, and so CN is fibrewise compact over B^2, since CN is closed; therefore $\mathscr{F}$ admits an adherence point $(\xi, \eta) \in CN \subset C\Delta$, where $\xi, \eta \in X_b$. Therefore $\xi \in \overline{\{\eta\}}$, by (13.5), and so $\xi \in N[\eta]$. Hence $(\xi, \eta) \in N$ and so we have our contradiction.

We continue the proof of (17.1) by showing that (12.1)–(12.3) are satisfied by the neighbourhoods of Δ. Of these conditions (12.1) and (12.2) are obvious. To establish (12.3) suppose, to obtain a contradiction, that there exists a point b of B and an open symmetric neighbourhood N of Δ, such that $(X_W^2 \cap M) \circ (X_W^2 \cap M) \subset N$ is false for each neighbourhood W of b and each neighbourhood M of Δ. Then the intersections

$$((X_W^2 \cap M) \circ (X_W^2 \cap M)) \cap CN$$

---

† To avoid the regularity assumption on the base space, in this proposition and its corollaries, it seems necessary to adopt a more elaborate notion of fibrewise uniform structure, as in [35].

generate a (b, b)-filter $\mathscr{F}$ on CN . By fibrewise compactness there exists an adherence point (ξ, η) of $\mathscr{F}$, where $\xi, \eta \in X_b$. Then $\xi \notin \overline{\{\eta\}}$ since $\eta \notin N[\xi]$.

Since X is fibrewise Hausdorff there exist disjoint open neighbourhoods U, V of ξ, η, respectively. Since X is fibrewise normal, by (3.25), there exist a neighbourhood W of b and closed neighbourhoods $X_W \cap H$, $X_W \cap K$ of ζ, η in X_W such that $X_W \cap H \subset U$, $X_W \cap K \subset V$. Write $L = X_W - X_W \cap (H \cup K)$ and consider the neighbourhood

$$M = (U \times U) \cup (V \times V) \cup (L \times L)$$

of ΔX_W. Now

$$(H \times X_W) \cap M = X_W^2 \cap (H \times U), \qquad (X_W \times K) \cap M = X_W^2 \cap (V \times K)$$

and so the neighbourhood $X_W^2 \cap (H \times K)$ of (ξ, η) in X_W^2 does not meet

$$M' = (X_W^2 \cap M) \circ (X_W^2 \cap M).$$

By regularity there exists a neighbourhood W' of b such that $W' \subset \bar{W}' \subset W$. Then $M' \cup C\Delta X_{W'}$ is a neighbourhood of ΔX and so $X_{W'}^2 \cap M' = X_{W'}^2 \cap (M' \cup C\Delta X_{W'})$ is a member of $\mathscr{F}$ which does not meet $X_{W'}^2 \cap (H \times K)$. This gives us our contradiction.

Clearly the fibrewise uniform topology cannot be coarser than the given fibrewise topology, since if $x \in X_b$, where $b \in B$, then $N[x]$ is a neighbourhood of x for each neighbourhood N of Δ. To prove the converse we first show that the closure of x in the fibrewise uniform topology is contained in the closure of x in the given fibrewise topology. To avoid confusion let us use the bar symbol only in the former sense, at this stage in the argument. Suppose that $\xi \notin \overline{\{x\}}$, where $\xi \in X_b$. Then $\overline{\{\xi\}}$ does not meet $\overline{\{x\}}$, by fibrewise regularity, and so $N = X^2 - (\{x\} \times \{\xi\})$ is a neighbourhood of Δ. Thus $\xi \notin N[x]$ and so ξ does not belong to the closure of x in the fibrewise uniform topology.

Finally let V be an open neighbourhood of x in the given fibrewise topology. Suppose, to obtain a contradiction, that $X_W \cap N[x] \not\subset V$ for all neighbourhoods W of b and all neighbourhoods N of Δ. Then the intersections $X_W \cap N[x] \cap CV$ generate a b-filter $\mathscr{F}$ on CV. Now CV is closed in X, which is fibrewise compact, and so CV is fibrewise compact. Therefore $\mathscr{F}$ has an adherence point $\xi \in CV$, where $\xi \in X_b$. But ξ belongs to the closure of x in the fibrewise uniform topology, and so to the closure of x in the given fibrewise topology. Hence $\xi \in V$, by fibrewise regularity. The contradiction shows that V is a neighbourhood of x in the fibrewise uniform topology, as well as the given fibrewise topology, and so we conclude that the fibrewise topologies coincide as required.

Corollary (17.2). *Let $\phi: X \to Y$ be a fibrewise function, where X and Y are fibrewise uniform over B, with B regular. Suppose that X is fibrewise compact in the fibrewise uniform topology. If ϕ is continuous, in the fibrewise uniform topology, then ϕ is fibrewise uniformly continuous.*

Corollary (17.3). *Let X be fibrewise locally compact and fibrewise Hausdorff over B, with B regular. Then X admits a fibrewise uniform structure compatible with the given fibrewise topology.*

For let X_B^+ be the fibrewise Alexandroff compactification of X, as in §8. Recall that X has the induced fibrewise topology with respect to the canonical fibrewise function $\rho: X \to X_B^+$. If X is fibrewise locally compact and fibrewise Hausdorff then X_B^+ is fibrewise Hausdorff, as well as fibrewise compact, and so is fibrewise uniformizable, by (17.1). Hence X itself is fibrewise uniformizable, through the induced fibrewise uniform structure.

Next we come to a series of results about fibrewise precompactness, beginning with the definition of the term.

Definition (17.4). *Let X be fibrewise uniform over B. Then X is fibrewise precompact (or fibrewise totally bounded) if the fibrewise separated completion $\hat{X}_B$ of X is fibrewise compact.*

Proposition (17.5). *Let $\phi: X \to Y$ be a fibrewise uniformly continuous surjection, where X and Y are fibrewise uniform over B. If X is fibrewise precompact then so is Y.*

For consider the diagram shown below, where ρ and σ are canonical.

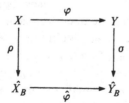

We have $\hat{\phi}\rho X = \sigma \phi X = \tilde{Y}_B$, since ϕ is surjective. Also the closure $\hat{X}_B$ of $\tilde{X}_B$ is fibrewise compact, by hypothesis, and so the closure of $\hat{\phi}\tilde{X}_B$ in $\hat{Y}_B$ is fibrewise compact. But the closure of $\hat{\phi}\tilde{X}_B$ is just the closure of $\tilde{Y}_B$, i.e. $\hat{Y}_B$ itself. This proves (17.5).

Proposition (17.6). *Let Z be a fibrewise set over B. Let $\{X_r\}$ be a family of fibrewise uniform spaces over B and let $\{\phi_r\}$ be a family of fibrewise* *functions, where* $\phi_r: Z \to X_r$. *Suppose that* Z *has the fibrewise uniform structure induced by* $\{\phi_r\}$. *Then a subset* A *of* Z *is fibrewise precompact if and only if the image* $\phi_r A$ *is precompact for each index* r.

The necessity of the condition is due to (17.5) while the sufficiency follows from (16.6) and from (4.4), the fibrewise Tychonoff theorem.

The following necessary and sufficient condition for fibrewise precompactness is illuminating.

Proposition (17.7). *Let* X *be fibrewise uniform over* B. *Then* X *is fibrewise precompact if and only if for each point* b *of* B *and each entourage* D *there exists a neighbourhood* W *of* b *such that* X_W *can be covered by a finite number of* D-small subsets.

For suppose that X is fibrewise precompact. Let D be a symmetric entourage of X. Then $(X_W^2 \cap E) \circ (X_W^2 \cap E) \subset D$, for some neighbourhood W of b and some symmetric entourage E, and so

$$((\hat{X}_B)_W^2 \cap E^*) \circ ((\hat{X}_B)_W^2 \cap E^*) \subset D^*,$$

as we have seen in the proof of (15.1). Since $\hat{X}_B$ is fibrewise compact there exists a neighbourhood $W' \subset W$ of b such that $(\hat{X}_B)_{W'}$ can be covered by a finite subfamily $\{V_j\}$ $(j = 1, \ldots, n)$ of the family $(\hat{X}_B)_{W'} \cap E^*[\xi]$ $(\xi \in (\hat{X}_B)_b)$. Then $\{U_j\}$ $(j = 1, \ldots, n)$ is a covering of $X_{W'}$, where $U_j = \rho^{-1} V_j$. But each of the U_j is $(X_{W'}^2 \cap E) \circ (X_{W'}^2 \cap E)$-small and so D-small. Since the symmetric entourages form a fibrewise basis this proves (17.7) in one direction.

For the proof in the other direction we suppose, for each point b of B and entourage D of X, that there exists a neighbourhood W of b for which X_W can be covered by a finite number of D-small sets. We shall show that this condition implies that every b-ultrafilter $\mathscr{F}$ on $\hat{X}_B$ is Cauchy and hence convergent, since $\hat{X}_B$ is fibrewise complete.

So let D^* be a fibrewise basic entourage of $\hat{X}_B$. I assert that the b-ultrafilter $\mathscr{F}$ contains a D^*-small member. By (13.5) there exists a neighbourhood W' of b and a closed symmetric entourage E^* of $\hat{X}_B$ such that $(\hat{X}_B)_{W'}^2 \cap E^* \subset D^*$. Consider the entourage $E = \rho^{-2} E^*$ of X. By hypothesis there exists a neighbourhood $W \subset W'$ of b, such that X_W can be covered by a finite number of E-small sets, say $U_1, \ldots, U_n$. Write $V_j = \rho U_j$ $(j = 1, \ldots, n)$. Now $(\hat{X}_B)_W$ is the closure of ρX_W and so is covered by $\{\bar{V}_j\}$ $(j = 1, \ldots, n)$. But $V_j \times V_j \subset E^*$, which is closed in $(\hat{X}_B)^2$, and so $\bar{V}_j \times \bar{V}_j \subset E^*$, i.e. the $\bar{V}_j$ are E^*-small. Also $(\hat{X}_B)_W$ is a member of $\mathscr{F}$, since $\mathscr{F}$ is a b-filter, and so one of the $\bar{V}_j$ is a member of $\mathscr{F}$, by (4.1), since $\mathscr{F}$

is a b-ultrafilter. This proves the assertion and shows that $\hat{X}_B$ is fibrewise compact, i.e. X is fibrewise precompact.

Corollary (17.8). *Let X be fibrewise uniform over B. If X can be covered by a finite number of fibrewise precompact subspaces then X is fibrewise precompact.*

Proposition (17.9). *Let $\phi \colon X \to Y$ be a fibrewise function, where X is a fibrewise set and Y is a fibrewise uniform space over B. Suppose that X has the fibrewise uniform structure induced from that of Y by means of ϕ. If Y is fibrewise precompact then so is X.*

For let E be an entourage of Y and let $b \in B$. If Y is fibrewise precompact then, by (17.7), there exist a neighbourhood W of b and a covering $A_1, \ldots, A_n$ of Y_W by E-small subsets of Y. Then there exist a neighbourhood $W' \subset W$ of b and an entourage D of X such that $X_{W'}^2 \cap D \subset \phi^{-2}E$, and $\phi^{-1}A_1, \ldots, \phi^{-1}A_n$ is a covering of $X_{W'}$ by $(X_{W'}^2 \cap D)$-small subsets of X. Since X has the induced fibrewise uniform structure this is sufficient to prove (17.9). Of course it follows at once, as a special case, that every subspace of a fibrewise precompact space is also fibrewise precompact.

Proposition (17.10). *Let X be fibrewise uniform over B, and let A be a dense subspace of X. If A is fibrewise precompact then so is X.*

For let D be an entourage of X and let $b \in B$. There exists a neighbourhood W' of b and a symmetric entourage E of X such that $(X_{W'}^2 \cap E) \circ (X_{W'}^2 \cap E) \subset D$. Consider the trace F of E on A^2. If A is fibrewise precompact there exists a neighbourhood $W \subset W'$ of b such that $A_W \subset F[S]$ for some finite subset S of A. I assert that $X_W \subset D[S]$. For if $x \in X_W$ the neighbourhood $X_W \cap E[x]$ of x has non-empty intersection with A and hence with A_W. Therefore $X_W \cap E[x]$ has non-empty intersection with $F[a]$, for some $a \in S$, and so

$$x \in ((X_W^2 \cap E) \circ (X_W^2 \cap E))[a].$$

Since $(X_W^2 \cap E) \circ (X_W^2 \cap E) \subset D$ this proves the assertion and hence (17.10).

Finally we prove

Proposition (17.11). *Let X be fibrewise uniform over B. Then X is fibrewise precompact if and only if each tied filter on X admits a Cauchy refinement.*

For let X be fibrewise precompact. Let D be an entourage of X and let $b \in B$. Then there exist a neighbourhood W' of b and a symmetric entourage E such that $(X_{W'}^2 \cap E) \circ (X_{W'}^2 \cap E) \subset D$, and there exists a neighbourhood $W \subset W'$ of b such that $X_W \subset E[S]$ for some finite subset S. Let $\mathcal{F}$ be a b-ultrafilter on X. Since X_W is a member of $\mathcal{F}$, so is $E[S]$. Therefore $X_W \cap E[x] \in \mathcal{F}$, by (4.1), for some $x \in S$. Since $X_W \cap E[x]$ is D-small this shows that $\mathcal{F}$ is Cauchy. Since each b-filter can be refined by a b-ultrafilter this proves the result in one direction.

To complete the proof, suppose that X is not fibrewise precompact. Then for some point b of B there exists an entourage D such that $X_W \subset D[S]$ is false for every neighbourhood W of b and every finite subset S of X. Consider the b-filter $\mathcal{F}$ on X which is generated by the family of subsets $X_W \cap (X - D[S])$, where W runs through the neighbourhoods of b and S runs through the finite subsets. I assert that $\mathcal{F}$ does not admit a Cauchy refinement.

For suppose, to obtain a contradiction, that there exists a Cauchy refinement $\mathcal{G}$ of $\mathcal{F}$. Then some D-small subset M of X is a member of $\mathcal{G}$. Now M and $X_W \cap (X - D[S])$ have non-empty intersection for each finite subset S of X. If x is a point of the intersection, for specific S, then $X_W \cap M$ is disjoint from $X - D[S \cup \{x\}]$ since $M \subset D[x]$. But $S \cup \{x\}$ is finite, since S is finite, and so we obtain a contradiction. This completes the proof of (17.11).

Exercises

1. Take $B = \{0, 1\}$, with the discrete topology, and take $X = \{0, 1, 2\}$, with the projection p given by $p(0) = 0$ and $p(1) = p(2) = 1$. Show that the subset $\Delta \cup \{(0, 1), (1, 0), (0, 2), (2, 0)\}$ of $X \times X$ generates a fibrewise uniform structure but not a uniform structure on X, and determine the fibrewise uniform topology.

2. Let X be a fibrewise uniform space over B. A *fibrewise uniform cover* of the fibre X_b, where $b \in B$, is a cover of the form $\{X_W \cap D[x] : x \in X_W\}$, where W is a neighbourhood of b and D is an entourage of X. Show that for any pair of fibrewise uniform covers of X_b there exists a fibrewise uniform cover of X_b which is a star-refinement of them both.

3. Let X be a fibrewise precompact fibrewise uniform space over B. Show that there exists a finite fibrewise uniform cover of each fibre X_b ($b \in B$).

4. Let $\phi: X \to Y$ be a fibrewise uniformly continuous function, where X and Y are fibrewise uniform spaces over B. Suppose that ϕ is *fibrewise uniformly open*, in the sense that for each entourage D of X and each point b of B there exists a neighbourhood W of b and an entourage E of Y such that

$$Y_W \cap E[\phi(x)] \subset \phi(D[x])$$

for all points $x \in X_W$. Show that ϕ is open, in the fibrewise uniform topology.

5. Let X be a fibrewise uniform space over B. Let R be a fibrewise equivalence relation on X. Suppose that for each entourage D of X and point b of B there exist a neighbourhood W of b and an entourage D' of X such that

$$X_W^2 \cap (R \circ D') \subset D \circ R.$$

Show that a fibrewise uniform structure on the fibrewise quotient set X/R can be defined so that the natural projection $\pi: X \to X/R$ is fibrewise uniformly open and fibrewise uniformly continuous. Also show that a fibrewise function $\phi: X/R \to Y$, where Y is fibrewise uniform, is fibrewise uniformly continuous if and only if the composition $\phi\pi: X \to Y$ is fibrewise uniformly continuous.

6. Let X be a fibrewise compact and fibrewise Hausdorff space over B. Let Y be a fibrewise complete and fibrewise separated fibrewise uniform space over B. Show that a fibrewise function $\phi: A \to Y$, where A is dense in X, can be extended to a continuous fibrewise function over the whole of X if and only if ϕ is fibrewise uniformly continuous.

7. Let X be fibrewise uniform over B. Let H be a closed and let K be a fibrewise compact subset of X. Suppose that $H \cap K$ does not meet X_b, where $b \in B$. Show that there exist a neighbourhood W of b and an entourage D of X such that $D[H] \cap D[K]$ does not meet X_W.

IV
Fibrewise homotopy theory

---

18. Fibrewise homotopy

Anyone who has worked on the theory of fibre bundles, and in related areas of mathematics, will have become aware of the need for a fibrewise version of homotopy theory. Isolated results can be found in the literature but nothing systematic. The present chapter is an attempt to provide an outline for such a theory. As we shall see, rather more is involved than simply taking one of the standard accounts of ordinary homotopy theory and inserting the word 'fibrewise' at every opportunity. However, I have kept as close as possible to the version of homotopy theory given in the latter part of [32], which is derived from the work of Puppe [54], Strøm [59], [60] and many others.

It seems reasonable at this stage to ease our terminology somewhat by writing fibrewise space instead of fibrewise topological space, and fibrewise map instead of continuous fibrewise function, except in the rare circumstances where this might cause some confusion.

Fibrewise homotopy can be defined in either of two equivalent ways. Specifically let us consider fibrewise maps of X into Y, where X and Y are fibrewise spaces over B. The first definition of fibrewise homotopy involves the (fibrewise) *cylinder*

$$I \times X = (I \times B) \times_B X$$

on the domain, which comes equipped with a family of fibrewise embeddings $\sigma_t \colon X \to I \times X$ $(0 \leqslant t \leqslant 1)$, where

$$\sigma_t(x) = (t, x) \quad (x \in X).$$

Definition (18.1). *Let $\theta, \phi \colon X \to Y$ be fibrewise maps, where X and Y are fibrewise spaces over B. A fibrewise homotopy of θ into ϕ is a fibrewise map $f \colon I \times X \to Y$ such that $f\sigma_0 = \theta$ and $f\sigma_1 = \phi$.*

Thus a fibrewise homotopy is just a homotopy in the ordinary sense which is a fibrewise map at each stage of the deformation.

The other definition of fibrewise homotopy involves the *fibrewise path-space*

$$P_B(Y) = \text{map}_B(I \times B, Y)$$

on the codomain, which comes equipped with a family of fibrewise projections $\rho_t \colon P_B(Y) \to Y$ $(0 \leqslant t \leqslant 1)$, where

$$\rho_t(\lambda) = \lambda(t) \quad (b \in B, \lambda \in P(Y_b));$$

here we interpret maps $I \times \{b\} \to Y_b$ as paths in Y_b in the obvious way, as in §9.

Definition (18.2). *Let* $\theta, \phi \colon X \to Y$ *be fibrewise maps, where X and Y are fibrewise spaces over B. A* fibrewise homotopy *of* θ *into* ϕ *is a fibrewise map* $\hat{f} \colon X \to P_B(Y)$ *such that* $\rho_0 \hat{f} = \theta$ *and* $\rho_1 \hat{f} = \phi$.

Thus if $f \colon I \times X \to Y$ is a fibrewise homotopy of θ into ϕ in the sense of (18.1) then the adjoint $\hat{f} \colon X \to P_B(Y)$ of f is a fibrewise homotopy of θ into ϕ in the sense of (18.2), and vice versa.

If there exists a fibrewise homotopy of θ into ϕ we say that θ is *fibrewise homotopic* to ϕ and write $\theta \simeq_B \phi$. A fibrewise homotopy into a fibrewise constant map is called a *fibrewise nulhomotopy*.

As in the ordinary theory one shows that fibrewise homotopy constitutes an equivalence relation between fibrewise maps. The set of equivalence classes of fibrewise maps from X to Y is denoted by $\pi_B(X, Y)$. Note that if $f_t \colon X \to Y$ is a fibrewise homotopy of θ into ϕ, then $\lambda^* f_t \colon \lambda^* X \to \lambda^* Y$ is a fibrewise homotopy of $\lambda^* \theta$ into $\lambda^* \phi$, for each space B' and map $\lambda \colon B' \to B$. Consequently λ determines a function

$$\lambda_\# \colon \text{map}_B(X, Y) \to \text{map}_{B'}(\lambda^* X, \lambda^* Y).$$

The operation of composition for fibrewise homotopy classes constitutes a function

$$\pi_B(Y, Z) \times \pi_B(X, Y) \to \pi_B(X, Z),$$

where X, Y, Z are fibrewise spaces over B. From the formal point of view π_B constitutes a binary functor from the category of fibrewise spaces to the category of sets, contravariant in the first entry and covariant in the second. There is a natural equivalence between the set $\pi_B(X, Y_1 \times_B Y_2)$ and the Cartesian product $\pi_B(X, Y_1) \times \pi_B(X, Y_2)$, for all fibrewise spaces X, Y_1 and Y_2, obtained by postcomposing with the projections. There is also a natural equivalence between the set $\pi_B(X_1 +_B X_2, Y)$ and the Cartesian product $\pi_B(X_1, Y) \times \pi_B(X_2, Y)$, for all fibrewise spaces X_1, X_2 and Y, obtained by precomposing with the insertions.

A fibrewise map $\phi: X \to Y$ is called a *fibrewise homotopy equivalence* if there exists a fibrewise map $\psi: Y \to X$ such that $\psi\phi \simeq_B \mathrm{id}_X$ and $\phi\psi \simeq_B \mathrm{id}_Y$; in that case ψ is said to be the *fibrewise homotopy inverse* of ϕ. The existence of a fibrewise homotopy equivalence determines an equivalence relation between fibrewise spaces: the equivalence classes are called *fibrewise homotopy types*.

In particular, fibrewise spaces which are of the same fibrewise homotopy type as the base space are said to be *fibrewise contractible*. Clearly a fibrewise space X is fibrewise contractible if and only if the identity on X is fibrewise nulhomotopic. A fibrewise space which is contractible, as an ordinary space, is not necessarily fibrewise contractible. For an example take the subspace $(I \times \{0\}) \cup (\{0\} \times I)$ of $I \times I$, regarded as a fibrewise space over I under the first projection.

There is a point which needs to be made here about the fibrewise quotient topology. Suppose that $\phi: X \to Y$ is a fibrewise quotient map. Then $\mathrm{id} \times \phi: I \times X \to I \times Y$ is also a fibrewise quotient map, by (5.3), since I is compact. For any fibrewise space Z, therefore, a fibrewise function $\psi: I \times Y \to Z$ is continuous if and only if the corresponding function $\psi(\mathrm{id} \times \phi): I \times X \to Z$ is continuous. In other words the family of fibrewise functions $\psi_t: Y \to Z$ $(t \in I)$ constitutes a fibrewise homotopy if and only if the family of fibrewise functions $\psi_t\phi: X \to Z$ constitutes a fibrewise homotopy.

In particular, suppose that $Y = X/R$, where R is a fibrewise equivalence relation on X. We have seen earlier that fibrewise maps of Y into Z correspond precisely to invariant fibrewise maps of X into Z. We now see that fibrewise homotopies of fibrewise maps of Y and Z correspond precisely to invariant fibrewise homotopies of invariant fibrewise maps of X into Z.

For example, consider the fibrewise cone $\Gamma_B(X)$ on the fibrewise space X. The fibrewise homotopy $f_t: I \times X \to I \times X$ $(t \in I)$ given by

$$f_t(s, x) = (st, x) \quad (s \in I, x \in X)$$

induces a fibrewise nulhomotopy of the identity on $\Gamma_B(X)$. Thus $\Gamma_B(X)$ is fibrewise contractible.

Let $\theta: X \to Y$ and $\phi: Y \to X$ be fibrewise maps, where X and Y are fibrewise spaces over B. Suppose that $\phi\theta$ is fibrewise homotopic to the identity on X. Then ϕ is said to be a *left inverse* of θ up to fibrewise homotopy, and θ is said to be a *right inverse* of ϕ up to fibrewise homotopy. Note that if θ admits both a left inverse ϕ and a right inverse ϕ', up to fibrewise homotopy, then $\phi \simeq_B \phi'$ and θ is a fibrewise homotopy equivalence.

Note that if X and Y are of the same fibrewise homotopy type over B then λ^*X and λ^*Y are of the same fibrewise homotopy type over B' for each space B' and map $\lambda: B' \to B$. Also note that the fibrewise homotopy types of the fibrewise topological product $X_1 \times_B X_2$ and coproduct $X_1 +_B X_2$ depend only on the fibrewise homotopy types of the fibrewise spaces X_1 and X_2.

Let X be a fibrewise space over B. If X has the same fibrewise homotopy type as $B \times T$, for some space T, then X is said to be *trivial*, with fibre T, in the sense of fibrewise homotopy type. *Local triviality*, in the same sense, is defined similarly.

Proposition (18.3). *Let X, Y, Z be fibrewise spaces over B. Then for any fibrewise map $\theta: X \to Y$ the fibrewise homotopy class of the precomposition function*

$$\theta^*: \mathrm{map}_B(Y, Z) \to \mathrm{map}_B(X, Z)$$

depends only on the fibrewise homotopy class of θ. Also for any fibrewise map $\phi: Y \to Z$ the fibrewise homotopy class of the post-composition function

$$\phi_*: \mathrm{map}_B(X, Y) \to \mathrm{map}_B(X, Z)$$

depends only on the fibrewise homotopy class of ϕ.

It follows that the fibrewise homotopy type of $\mathrm{map}_B(X, Y)$ depends only on the fibrewise homotopy types of X and Y. In particular $\mathrm{map}_B(X, Y)$ has the same fibrewise homotopy type as Y whenever X is fibrewise contractible, while $\mathrm{map}_B(X, Y)$ is fibrewise contractible whenever Y is fibrewise contractible.

To prove the first assertion in (18.3) consider the fibrewise function

$$\lambda: I \times \mathrm{map}_B(I \times X, Z) \to \mathrm{map}_B(X, Z)$$

which is given over each point b of B by

$$\lambda(t, h)(x) = h(t, x) \quad (t \in I, x \in X_b, h: I \times X_b \to Z_b).$$

Let $(K, V; W)$ be a fibrewise compact-open subset of $\mathrm{map}_B(X, Z)$, where W is a neighbourhood of b, where $V \subset Z_W$ is open and where $K \subset X_W$ is fibrewise compact over W. Let $\beta \in W$ and let $(t, k) \in \lambda^{-1}(K, V; W)$, where $t \in I$ and $k: I \times X_\beta \to Z_\beta$. Then we have $k(\{t\} \times K_\beta) \subset V_\beta$. Since K is fibrewise compact over W there exists a neighbourhood $W' \subset W$ of β and a neighbourhood N of t such that $k(N \times K_{W'}) \subset V_{W'}$. Since I is compact and regular we can shrink N to a neighbourhood U of t such that $\bar{U}$ is compact. Then $U \times (\bar{U} \times K_{W'}, V_{W'}; W')$ is a neighbourhood of (t, k) which is contained in $\lambda^{-1}(K, V; W)$. Thus λ is continuous and the first part of (18.3) follows by precomposing with a fibrewise homotopy $I \times X \to Y$.

To prove the second part, consider the fibrewise function

$$\mu: I \times \text{map}_B(X, Y) \to \text{map}_B(X, I \times Y)$$

which is given over each point b of B by

$$\mu(t, f)(x) = (t, f(x)) \quad (t \in I, x \in X, f: X_b \to Y_b).$$

Let $(K, N \times V; W)$ be a fibrewise compact-open subset of $\text{map}_B(X, I \times Y)$, where W is a neighbourhood of b, where $N \subset I$ and $V \subset Y_W$ are open, and where $K \subset X_W$ is fibrewise compact over W. Let $\beta \in W$ and let $(t, g) \in (K, N \times V; W)$, where $t \in I$ and $g: X_\beta \to Y_\beta$. Then $t \in N$ and $gK_\beta \subset V_\beta$. Hence $N \times (K, V; W)$ is a neighbourhood of (t, g) which is contained in $\mu^{-1}(K, N \times V; W)$. It follows, using (9.13), that μ is continuous, and so the second part of (18.3) is obtained by post-composing μ with a fibrewise homotopy $I \times Y \to Z$.

Let $u: A \to X$ be a fibrewise map, where A and X are fibrewise spaces over B. Let $\theta, \phi: X \to Y$ be fibrewise maps, where Y is a fibrewise space, such that $\theta u = \phi u$. By a *fibrewise homotopy* of θ into ϕ *under A* we mean a fibrewise homotopy f_t of θ into ϕ such that $f_t u$ is independent of t. When such a fibrewise homotopy exists we say that θ and ϕ are fibrewise homotopic *under A* and write $\theta \simeq_B^A \phi$. Again we have an equivalence relation, and the set of equivalence classes is denoted by $\pi_B^A(X, Y)$. Of course, $\pi_B^A(X, Y)$ reduces to $\pi_B(X, Y)$ when A is empty.

In the case in which u is an embedding, in particular when A is a subspace of X with u the inclusion, it is usual to describe a fibrewise homotopy under A as *relative* to A, or rel A for short.

Definition (18.4). *Let X be a fibrewise space over B. A subspace A of X is a fibrewise deformation retract if there exists a fibrewise homotopy $f_t: X \to X$ rel A of the identity such that $f_1 X = A$.*

In that case f_1 is a fibrewise retraction, and the inclusion $u: A \to X$ is a fibrewise homotopy equivalence with inverse f_1, up to fibrewise homotopy.

For example, let $\phi: X \to Y$ be a fibrewise map, where X and Y are fibrewise spaces over B. The fibrewise push-out of the cotriad

$$I \times X \xleftarrow{\sigma_0} X \xrightarrow{\phi} Y$$

is called the *fibrewise mapping cylinder* of ϕ and is written $M_B(\phi)$. Thus

$M = M_B(\phi)$ comes equipped with fibrewise maps

$$I \times X \to M \leftarrow Y;$$

the latter is an embedding. Moreover a fibrewise map $\rho: M \to Y$ is given by $\phi \pi_2$ on $I \times X$ and by the identity on Y. We can regard M, with section given by σ_0, as the fibrewise cone on X, considered as a fibrewise space over Y through ϕ, and so M is fibrewise contractible, in this sense. Hence it follows that Y, as a fibrewise space over B, is a fibrewise deformation retract of M.

We conclude this section by proving another result about fibrewise deformation retracts of which the full significance will not emerge until §20.

Proposition (18.5). *Let (X, A) be a fibrewise pair over B. If $(\{0\} \times X) \cup (I \times A)$ is a fibrewise retract of $I \times X$ then $(\{0\} \times X) \cup (I \times A)$ is a fibrewise deformation retract of $I \times X$.*

For let $r: I \times X \to (\{0\} \times X) \cup (I \times A)$ be a fibrewise retraction, and let

$$I \underset{r_1}{\leftarrow} I \times X \underset{r_2}{\to} X$$

be the components of r. We define $R: I \times I \times X \to I \times X$ by

$$R(s, t, x) = ((1 - s)t + sr_1(t, x), r_2(st, x)),$$

where $s, t \in I$ and $x \in X$. Then R is a fibrewise homotopy of the identity on $I \times X$ into r, leaving $(\{0\} \times X) \cup (I \times A)$ fixed.

19. Fibrewise pointed homotopy

We now turn our attention again to the category of fibrewise pointed spaces. Fibrewise pointed homotopy is an equivalence relation between fibrewise pointed maps which can be defined in either of two equivalent ways, just as in the non-pointed case.

Specifically let us consider fibrewise pointed maps of X into Y, where X and Y are fibrewise pointed spaces over B. The first definition of fibrewise pointed homotopy involves the reduced fibrewise cylinder $I \tilde{\times} X$ on the domain, i.e. the fibrewise push-out of the cotriad

$$I \times X \underset{\mathrm{id} \times s}{\leftarrow} I \times B \underset{\pi_2}{\to} B,$$

where s is the section. This comes equipped with a family of fibrewise pointed embeddings $\sigma_t: X \to I \tilde{\times} X$ $(0 \leqslant t \leqslant 1)$, induced by the corresponding functions in the non-pointed case.

Definition (19.1). *Let* θ, $\phi\colon X \to Y$ *be fibrewise pointed maps, where X and Y are fibrewise pointed spaces over B. A* fibrewise pointed homotopy *of θ into ϕ is a fibrewise pointed map $f\colon I \tilde{\times} X \to Y$ such that $f\sigma_0 = \theta$ and $f\sigma_1 = \phi$.*

Thus a fibrewise pointed homotopy is induced by a fibrewise homotopy under B, where B is embedded in X by means of the section, and we shall often take this point of view.

The other definition of fibrewise pointed homotopy involves the *fibrewise pointed path-space*

$$P_B^B(Y) = \mathrm{map}_B^B(I \tilde{\times} B, Y)$$

on the codomain, which comes equipped with a family of fibrewise pointed projections $\rho_t\colon P_B^B(Y) \to Y$ $(0 \leqslant t \leqslant 1)$, given by the same formula as in the non-pointed case.

Definition (19.2). *Let* θ, $\phi\colon X \to Y$ *be fibrewise pointed maps, where X and Y are fibrewise pointed spaces over B. A* fibrewise pointed homotopy *of θ into ϕ is a fibrewise pointed map $\hat{f}\colon X \to P_B^B(Y)$ such that $\rho_0\hat{f} = \theta$ and $\rho_1\hat{f} = \phi$.*

Thus if $f\colon I \tilde{\times} X \to Y$ is a fibrewise pointed homotopy of θ into ϕ in the sense of (19.1) then the adjoint $\hat{f}\colon X \to P_B^B(Y)$ of f is a fibrewise pointed homotopy of θ into ϕ in the sense of (19.2), and vice versa.

If there exists a fibrewise pointed homotopy of θ into ϕ we say that θ is *fibrewise pointed homotopic* to ϕ and write $\theta \simeq_B^B \phi$. A fibrewise pointed homotopy into the fibrewise constant map is called a *fibrewise pointed nulhomotopy*.

As in the ordinary theory one shows that fibrewise pointed homotopy constitutes an equivalence relation between fibrewise pointed maps. The pointed set of equivalence classes of fibrewise pointed maps from X to Y is denoted by $\pi_B^B(X, Y)$. Note that if $f_t\colon X \to Y$ is a fibrewise pointed homotopy of θ into ϕ then $\lambda^* f_t\colon \lambda^* X \to \lambda^* Y$ is a fibrewise pointed homotopy of $\lambda^* \theta$ into $\lambda^* \phi$, for each space B' and map $\lambda\colon B' \to B$. Consequently λ determines a pointed function

$$\lambda_\#\colon \pi_B^B(X, Y) \to \pi_{B'}^{B'}(\lambda^* X, \lambda^* Y).$$

The operation of composition for fibrewise pointed homotopy classes constitutes a function

$$\pi_B^B(Y, Z) \wedge \pi_B^B(X, Y) \to \pi_B^B(X, Z),$$

where X, Y, Z are fibrewise pointed spaces over B. From the formal point of view π_B^B constitutes a binary functor from the category of fibrewise pointed spaces to the category of pointed sets, contravariant in the first entry and covariant in the second.

A natural equivalence between the pointed set $\pi_B^B(X, Y_1 \times_B Y_2)$ and the product $\pi_B^B(X, Y_1) \times \pi_B^B(X, Y_2)$, for all fibrewise pointed spaces X, Y_1 and Y_2, is obtained by postcomposing with the projections. Also a natural equivalence between the pointed set $\pi_B^B(X_1 \vee_B X_2, Y)$ and the product $\pi_B^B(X_1, Y) \times \pi_B^B(X_2, Y)$, for all fibrewise pointed spaces X_1, X_2 and Y, is obtained by precomposing with the insertions.

A fibrewise pointed map $\phi \colon X \to Y$ is called a *fibrewise pointed homotopy equivalence* if there exists a fibrewise pointed map $\psi \colon Y \to X$ such that $\psi\phi \simeq_B^B \mathrm{id}_X$ and $\phi\psi \simeq_B^B \mathrm{id}_Y$; in that case ψ is said to be the *fibrewise pointed homotopy inverse* of ϕ. The existence of a fibrewise pointed homotopy equivalence determines an equivalence relation between fibrewise pointed spaces; the equivalence classes are called *fibrewise pointed homotopy types*.

In particular, fibrewise pointed spaces which are of the same fibrewise pointed homotopy type as the base space are said to be *fibrewise pointed contractible*. Clearly a fibrewise pointed space X is fibrewise pointed contractible if and only if the identity on X is fibrewise pointed nulhomotopic. For example, if Y is a fibrewise pointed space over B the reduced fibrewise cone $\Gamma_B^B(Y)$ is fibrewise pointed contractible.

Let $\theta \colon X \to Y$ and $\phi \colon Y \to X$ be fibrewise pointed maps, where X and Y are fibrewise pointed spaces over B. Suppose that $\phi\theta$ is fibrewise pointed homotopic to the identity on X. Then ϕ is said to be a *left inverse* of θ up to fibrewise pointed homotopy, and θ is said to be a *right inverse* of ϕ up to fibrewise pointed homotopy. Note that if θ admits both a left inverse ϕ and a right inverse ϕ', up to fibrewise pointed homotopy, then $\phi \simeq_B^B \phi'$ and θ is a fibrewise pointed homotopy equivalence.

Note that if X and Y are of the same fibrewise pointed homotopy type over B then $\lambda^* X$ and $\lambda^* Y$ are of the same fibrewise pointed homotopy type over B', for each space B' and map $\lambda \colon B' \to B$. Also note that the fibrewise pointed homotopy types of the fibrewise topological product $X_1 \times_B X_2$, the fibrewise pointed coproduct $X_1 \vee_B X_2$ and the fibrewise smash product $X_1 \wedge_B X_2$ depend only on the fibrewise pointed homotopy types of the fibrewise pointed spaces X_1 and X_2.

Let X be a fibrewise pointed space over B. If X has the same fibrewise pointed homotopy type as $B \times T$, for some pointed space T, then X is said to be *trivial*, with fibre T, in the sense of fibrewise pointed homotopy type. *Local triviality*, in the same sense, is defined similarly.

Suppose that X and Y are fibrewise spaces over B, with subspaces $X' \subset X$ and $Y' \subset Y$. Fibrewise maps $(X, X') \to (Y, Y')$ of the pair induce fibrewise pointed maps $X/_B X' \to Y/_B Y'$ after fibrewise collapsing, and similarly with fibrewise homotopies. Hence fibrewise homotopy equivalences of the pair induce fibrewise pointed homotopy equivalences.

Proposition (19.3). *Let X, Y, Z be fibrewise pointed spaces over B. Then for any fibrewise pointed map $\theta\colon X \to Y$ the fibrewise pointed homotopy class of the precomposition function*

$$\theta^*\colon \operatorname{map}_B^B(Y, Z) \to \operatorname{map}_B^B(X, Z)$$

depends only on the fibrewise pointed homotopy class of θ. Also for any fibrewise pointed map $\phi\colon Y \to Z$ the fibrewise pointed homotopy class of the postcomposition function

$$\phi_*\colon \operatorname{map}_B^B(X, Y) \to \operatorname{map}_B^B(X, Z)$$

depends only on the fibrewise pointed homotopy class of ϕ.

The proof is very similar to that of (18.3) and will therefore be omitted. The result implies that the fibrewise pointed homotopy type of $\operatorname{map}_B^B(X, Y)$ depends only on the fibrewise pointed homotopy types of X and Y. In particular $\operatorname{map}_B^B(X, Y)$ is fibrewise pointed contractible whenever X or Y is fibrewise pointed contractible.

Given a fibrewise pointed space X over B we say that a subspace A of X is a *fibrewise pointed deformation retract* of X if the identity on X is fibrewise pointed homotopic, rel A, to a fibrewise pointed map with values in A. Note that a fibrewise homotopy rel A is necessarily fibrewise pointed.

For example, let $\phi\colon X \to Y$ be a fibrewise pointed map, where X and Y are fibrewise pointed spaces over B. The fibrewise push-out of the cotriad

$$I \tilde{\times} X \xleftarrow{\sigma_0} X \xrightarrow{\phi} Y$$

is called the *reduced fibrewise mapping cylinder* of ϕ and is written $M_B^B(\phi)$. Alternatively $M_B^B(\phi)$ may be obtained from the fibrewise mapping cylinder $M_B(\phi)$ by fibrewise collapsing the subspace $M_B(\operatorname{id}_B) = I \times B$. Thus $\tilde{M} = M_B^B(\phi)$ comes equipped with fibrewise pointed maps

$$I \tilde{\times} X \to \tilde{M} \leftarrow Y;$$

the latter is an embedding. Moreover, a fibrewise pointed map $\rho\colon \tilde{M} \to Y$ is given by $\phi\pi_2$ on $I \tilde{\times} X$ and by the identity on Y. We can regard $\tilde{M}$, with the above section, as the reduced fibrewise cone on X, considered as a fibrewise pointed space over Y through ρ, and so $\tilde{M}$ is fibrewise pointed contractible, in this sense. Hence it follows that Y is a fibrewise deformation retract of $\tilde{M}$, as a fibrewise pointed space over B.

The proof of the following result is similar to the proof of the corresponding result (18.5) in the previous section and will therefore be omitted.

Proposition (19.4). *Let* (X, A) *be a fibrewise pointed pair over B. If* $(\{0\} \times X) \cup (I \tilde{\times} A)$ *is a fibrewise retract of* $I \tilde{\times} X$ *then* $(\{0\} \times X) \cup (I \tilde{\times} A)$ *is a fibrewise pointed deformation retract of* $I \tilde{\times} X$.

Let us now turn our attention to fibrewise binary systems consisting of a fibrewise pointed space X over B and a fibrewise pointed map $m \colon X \times_B X \to X$, called the *fibrewise multiplication*. Among the conditions which may be satisfied by such a system the following turn out to be important.

Definition (19.5). *The fibrewise multiplication m is* fibrewise homotopy-commutative *if* $m \simeq_B^B mt$, *where t is the switching equivalence shown below.*

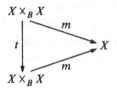

Definition (19.6). *The fibrewise multiplication m is* fibrewise homotopy-associative *if*

$$m(m \times \text{id}) \simeq_B^B m(\text{id} \times m),$$

where

$$X \times_B X \times_B X \xrightarrow[m \times \text{id}]{\text{id} \times m} X \times_B X \xrightarrow{m} X.$$

Definition (19.7). *The fibrewise multiplication m satisfies the* Hopf condition *if*

$$m(\text{id} \times c)\Delta \simeq_B^B \text{id} \simeq_B^B m(c \times \text{id})\Delta,$$

where

$$X \xrightarrow{\Delta} X \times_B X \xrightarrow[c \times \text{id}]{\text{id} \times c} X \times_B X \xrightarrow{m} X.$$

Here c, as usual, denotes the fibrewise constant map. A fibrewise binary system where the fibrewise multiplication satisfies the Hopf condition is called a *fibrewise Hopf space*.

Definition (19.8). *The fibrewise pointed map* $v: X \to X$ *is a* fibrewise homotopy-inversion *for the fibrewise multiplication m if*

$$m(\text{id} \times v)\Delta \simeq_B^B c \simeq_B^B m(v \times \text{id})\Delta,$$

where

$$X \xrightarrow{\Delta} X \times_B X \xrightarrow[v \times \text{id}]{\text{id} \times v} X \times_B X \xrightarrow{m} X.$$

A fibrewise homotopy-associative fibrewise Hopf space for which the fibrewise multiplication admits a fibrewise homotopy-inversion is called a *fibrewise group-like space*: in that case the fibrewise pointed homotopy class of the fibrewise homotopy-inversion is uniquely determined. For example, if T is a group-like space then $B \times T$ (with the obvious structure) is a fibrewise group-like space over B.

Obviously a fibrewise multiplication on the fibrewise pointed space X over B determines a multiplication on the pointed set $\pi_B^B(A, X)$ for all fibrewise pointed spaces A. If the former is fibrewise homotopy-commutative then the latter is commutative, and similarly with the other conditions. Thus $\pi_B^B(A, X)$ is a group if X is fibrewise group-like.

Similar results hold for fibrewise pointed mapping-spaces. Thus a fibrewise multiplication on X determines a fibrewise multiplication on $\text{map}_B^B(A, X)$ for all fibrewise regular pointed A. If the former is fibrewise homotopy-commutative then so is the latter, and similarly with the other conditions. Thus $\text{map}_B^B(A, X)$ is fibrewise group-like if X is fibrewise group-like.

Dually let us also consider fibrewise cobinary systems consisting of a fibrewise pointed space X over B and a fibrewise pointed map $m: X \to X \vee_B X$, called the *fibrewise comultiplication*.

Definition (19.9). *The fibrewise comultiplication m is* fibrewise homotopy-commutative *if* $m \simeq_B^B tm$, *where t is the switching equivalence as shown below.*

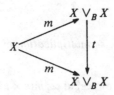

Definition (19.10). *The fibrewise comultiplication m is* fibrewise homotopy-associative *if*

$$(m \vee \text{id})m \simeq_B^B (\text{id} \vee m)m,$$

where

$$X \xrightarrow{m} X \vee_B X \xrightarrow[m \vee \text{id}]{\text{id} \vee m} X \vee_B X \vee_B X.$$

Definition (19.11). *The fibrewise comultiplication m satisfies the* coHopf *condition if*

$$\nabla(\mathrm{id} \vee c)m \simeq_B^B \mathrm{id} \simeq_B^B \nabla(c \vee \mathrm{id})m,$$

where

$$X \xrightarrow{m} X \vee_B X \underset{c \vee \mathrm{id}}{\overset{\mathrm{id} \vee c}{\longrightarrow}} X \vee_B X \xrightarrow{\nabla} X.$$

A fibrewise cobinary system where the fibrewise comultiplication satisfies the coHopf condition is called a *fibrewise coHopf space*.

Definition (19.12). *The fibrewise pointed map* $v: X \to X$ *is a* fibrewise homotopy-inversion *for the fibrewise comultiplication m if*

$$\nabla(\mathrm{id} \vee v)m \simeq_B^B c \simeq_B^B \nabla(v \vee \mathrm{id})m,$$

where

$$X \xrightarrow{m} X \vee_B X \underset{v \vee \mathrm{id}}{\overset{\mathrm{id} \vee v}{\longrightarrow}} X \vee_B X \xrightarrow{\nabla} X.$$

A fibrewise homotopy-associative fibrewise coHopf space which admits a fibrewise homotopy-inversion is called a *fibrewise cogroup-like space*; in that case the fibrewise pointed homotopy class of the fibrewise homotopy-inversion is uniquely determined. For example, if T is a cogroup-like space then $B \times T$ (with the obvious structure) is a fibrewise cogroup-like space over B. In particular $B \times S^1$ is fibrewise cogroup-like.

Obviously a fibrewise comultiplication on the fibrewise pointed space X over B determines a multiplication on the pointed set $\pi_B^B(X, Y)$, for all fibrewise pointed topological Y. If the former is fibrewise homotopy-commutative then the latter is commutative, and similarly with the other conditions. Thus $\pi_B^B(X, Y)$ is a group when X is fibrewise cogroup-like. This is the case (see below), when X is given as a reduced fibrewise suspension. If X is a fibrewise coHopf space and Y is a fibrewise Hopf space then the multiplication on $\pi_B^B(X, Y)$ determined by the fibrewise comultiplication on X coincides with the multiplication determined by the fibrewise multiplication on Y, and is both commutative and associative.

Similar results hold for pointed fibrewise mapping-spaces. Thus if X has closed section a fibrewise comultiplication on X determines a fibrewise multiplication on $\mathrm{map}_B^B(X, Y)$ for all fibrewise pointed topological spaces Y. If the former is fibrewise homotopy-commutative then so is the latter, and similarly with the other conditions. Thus $\mathrm{map}_B^B(X, Y)$ is fibrewise group-like if X is fibrewise cogroup-like. In particular the *fibrewise loop-space* $\Omega_B^B(Y) = \mathrm{map}_B^B(B \times S^1, Y)$ is fibrewise group-like for all fibrewise pointed Y. If X is a fibrewise coHopf space with closed section and Y is a fibrewise Hopf space the fibrewise multiplication on $\mathrm{map}_B^B(X, Y)$ determined by the fibrewise comultiplication on X is fibrewise pointed homotopic to the fibrewise multiplication determined by the fibrewise multiplication on Y, and is both fibrewise homotopy-commutative and fibrewise homotopy-associative.

If X has a closed section then, by (6.1), a fibrewise comultiplication on X determines a fibrewise comultiplication on $X \wedge_B Y$ for all fibrewise pointed spaces Y with closed section. If the former is fibrewise homotopy-commutative then so is the latter, and similarly with the other conditions. Thus $X \wedge_B Y$ is fibrewise cogroup-like if X is fibrewise cogroup-like; in particular the reduced fibrewise suspension $\Sigma_B^B(Y)$ is fibrewise cogroup-like, for all fibrewise pointed spaces Y with closed section.

Note that the adjoint, in the sense of §9, of a fibrewise pointed map $\Sigma_B^B(X) \to Y$ is a fibrewise pointed map $X \to \Omega_B^B(Y)$, for all fibrewise pointed X and Y. In this way a homomorphism

$$\pi_B^B(\Sigma_B^B(X), Y) \to \pi_B^B(X, \Omega_B^B(Y))$$

is defined. It follows from (9.13) that the homomorphism is an isomorphism whenever X is fibrewise compact and fibrewise regular.

20. Fibrewise cofibrations

Problems concerning the fibrewise extensions of fibrewise maps arise in various ways but the basic question is always of the same form. Thus let X be a fibrewise space over B and let A be a subspace of X. Let $f: A \to E$ be a fibrewise map, where E is a fibrewise space over B. The question is: does there exist a fibrewise extension of f to X, i.e. a fibrewise map $g: X \to E$ such that $g \mid A = f$?

For example, take $E = A$ with f the identity; does there exist a fibrewise retraction $X \to A$? For another example, take $X = I \times Y$ and $A = \dot{I} \times Y$, for some fibrewise space Y: given fibrewise maps $\theta, \phi: Y \to E$ does there exist a fibrewise homotopy of θ into ϕ?

Consideration of such problems is greatly facilitated if the inclusion $A \to X$ satisfies the following condition.

Definition (20.1). *The fibrewise map* $u: A \to X$, *where A and X are fibrewise spaces over B, is a fibrewise cofibration if u has the following fibrewise homotopy extension property. Let $f: X \to E$ be a fibrewise map, where E is a fibrewise space, and let $g: A \to P_B(E)$ be a fibrewise homotopy* *such that* $\rho_0 g = fu$. *Then there exists a fibrewise homotopy* $h\colon X \to P_B(E)$ *such that* $\rho_0 h = f$ *and* $hu = g$.

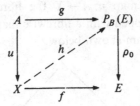

Instead of saying that u is a fibrewise cofibration we may sometimes say that X is a *fibrewise cofibre space under A*: for example the (fibrewise) coproduct of fibrewise cofibre spaces under A is a fibrewise cofibre space under A. An important special case is when A is a subspace of X. In that case we describe (X, A) as a *fibrewise cofibred pair* when the inclusion $A \to X$ is a fibrewise cofibration. For example (X, X) and $(X, \varnothing)$ are fibrewise cofibred pairs, for all fibrewise spaces X.

By taking E in the definition to be a product $B \times T$, for any space T, we see that a fibrewise cofibration $u\colon A \to X$ is necessarily a cofibration in the ordinary sense, in particular u is necessarily injective. The fibrewise push-out of the cotriad

$$X \xleftarrow{u} A \xrightarrow{p|A} B$$

is called the *fibrewise cofibre* of u. When $A \subset X$ and u is the inclusion the fibrewise cofibre is just the fibrewise collapse $X/_B A$.

Proposition (20.2). *Let* $u\colon A \to X$ *be a fibrewise cofibration, where A and X are fibrewise spaces over B. Then the push-out* $v\colon A' \to \xi_* X$ *is a fibrewise cofibration for each space A' over B and fibrewise map* $\xi\colon A \to A'$.

The proof is entirely formal and will be left to serve as an exercise.

Given a fibrewise map $u\colon A \to X$, where A and X are spaces over B, the fibrewise mapping cylinder $M = M_B(u)$ is defined as before to be the fibrewise push-out of the cotriad

$$I \times A \xleftarrow{\sigma_0} A \xrightarrow{u} X.$$

Thus M comes equipped with canonical fibrewise maps

$$I \times A \xrightarrow{H} M \xleftarrow{j} X$$

where j is a fibrewise homotopy equivalence, as we have seen. In the case in which A is a subspace of X, with u the inclusion, we have a continuous fibrewise bijection determined by the triad

$$I \times A \to (\{0\} \times X) \cup (I \times A) \leftarrow \{0\} \times X \equiv X.$$

If, moreover, A is closed in X then the correspondence is a fibrewise topological equivalence and we may identify M with $(\{0\} \times X) \cup (I \times A)$.

For each fibrewise map $u: A \to X$ the fibrewise mapping cylinder $M = M_B(u)$ comes equipped with a fibrewise map $k: M \to I \times X$ which is derived from the diagram shown below.

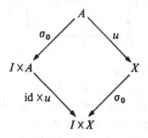

Proposition (20.3). *The fibrewise map* $u: A \to X$, *where* A *and* X *are fibrewise spaces over* B, *is a fibrewise cofibration if and only if the fibrewise map* k *admits a left inverse* $l: I \times X \to M$.

For suppose that l exists. Given a space E over B and fibrewise maps $f: X \to E$, $g: A \to P_B(E)$, related as in (20.1), we consider the triad

$$I \times A \xrightarrow{\hat{g}} E \xleftarrow{f} X,$$

where $\hat{g}$ is the adjoint of g. Precomposition of l with the fibrewise pull-back $M \to E$ of the triad yields a fibrewise map $I \times X \to E$, of which the adjoint $X \to P_B(E)$ provides a filler as required.

Conversely suppose that u is a fibrewise cofibration. Take E to be the fibrewise mapping cylinder $M = M_B(u)$ and consider the following diagram

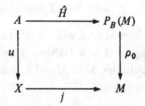

The adjoint of the filler $X \to P_B(M)$ is the required left inverse of k.

Corollary (20.4). *Let* $u: A \to X$ *be a fibrewise cofibration, where* A *and* X *are fibrewise spaces over* B. *Then* $u: A \to X$ *is a fibrewise cofibration over* B' *for each space* B' *and map* $B \to B'$.

From now on we shall concentrate, for simplicity, on closed fibrewise pairs (X, A), i.e. on fibrewise pairs (X, A) such that A is closed in X. In that case the fibrewise mapping cylinder can be replaced by $(\{0\} \times X) \cup (I \times A)$, as we saw in §18, and then (20.3) yields

Proposition (20.5). *Let (X, A) be a closed fibrewise pair over B. Then (X, A) is fibrewise cofibred if and only if $(\{0\} \times X) \cup (I \times A)$ is a fibrewise retract of $I \times X$.*

Corollary (20.6). *Let (X, A) be a closed fibrewise pair over B. If (X, A) is fibrewise cofibred over B then the pull-back (λ^*X, λ^*A) is fibrewise cofibred over B' for each space B' and map $\lambda: B' \to B$.*

Corollary (20.7). *Let (X, A) be a closed fibrewise pair over B. If (X, A) is fibrewise cofibred then so is*

$$T \times_B (X, A) = (T \times_B X, T \times_B A)$$

for all fibrewise spaces T.

Proposition (20.8). *Let (X, A) be a closed fibrewise cofibred pair over B. Suppose that A is fibrewise contractible. Then the natural projection $X \to X/_B A$ to the fibrewise collapse is a fibrewise homotopy equivalence.*

For let $f_t: A \to A$ be a fibrewise nulhomotopy of the identity, so that f_1 is fibrewise constant. The inclusion $g: A \to X$ of f_t can be fibrewise extended to a fibrewise homotopy $h_t: X \to X$ of the identity on X. Now h_1 induces a fibrewise map $h_1' = h_1 \pi^{-1}: X/_B A \to X$, where $\pi: X \to X/_B A$ denotes the natural projection. Similarly πh_t induces a fibrewise homotopy $h_t'': X/_B A \to X/_B A$, with $h_1'' = \pi h_1'$. Since $h_1' \pi = h_1 \simeq_B h_0 = \mathrm{id}_X$ and $\pi h_1' = h_1'' \simeq_B h_0'' = \mathrm{id}_{X/_B A}$ we conclude that h_1' is a fibrewise homotopy inverse of π, thus proving the result.

Proposition (20.9). *Let $u: A \to X$ and $v: A \to Y$ be fibrewise maps, where A, X and Y are fibrewise spaces over B. Let $\phi: X \to Y$ be a fibrewise map such that $\phi u \simeq_B v$. If u is a fibrewise cofibration then $\phi \simeq_B \psi$ for some fibrewise map $\psi: X \to Y$ such that $\psi u = v$.*

For let $g: I \times A \to Y$ be a fibrewise homotopy of ϕu into v. Since $g_0 = \phi u$ and since u is a fibrewise cofibration there exists a fibrewise extension $f: I \times X \to Y$ of g such that $f_0 \simeq_B^A \phi$. Take ψ to be f_1; then $\psi u = v$ as required.

Corollary (20.10). *Let $u: A \to X$ be a fibrewise cofibration, where A and X are fibrewise spaces over B. If u admits a left inverse up to fibrewise homotopy then u admits a left inverse.*

Proposition (20.11). *Let $u: A \to X$ be a fibrewise cofibration, where A and X are fibrewise spaces over B. Let $\theta: X \to X$ be a fibrewise map under A such that $\theta \simeq_B \mathrm{id}_X$. Then there exists a fibrewise map $\theta': X \to X$ under A such that $\theta'\theta \simeq_B^A \mathrm{id}_X$.*

For let f_t be a fibrewise homotopy of θ into id_X. Then $f_t u$ is a fibrewise homotopy of u into itself. Since u is a fibrewise cofibration there exists a fibrewise homotopy $g_t: X \to X$ of id_X such that $f_t u = g_t u$. Take $\theta' = g_1$; we shall prove that $\theta'\theta \simeq_B^A \mathrm{id}_X$. For consider the juxtaposition k_s of $g_{1-s}\theta$ and f_s, as a fibrewise homotopy of $\theta'\theta$ into id_X. Now $f_t u = g_t u$ and hence $H_{(s,0)} = k_s u$, where $H: I \times I \times A \to X$ is given by

$$H(s, t, a) = \begin{cases} g(1 - 2s(1 - t), u(a)) & (s \leqslant \tfrac{1}{2}) \\ f(1 - 2(1 - s)(1 - t), u(a)) & (s \geqslant \tfrac{1}{2}). \end{cases}$$

Since u is a fibrewise cofibration we can extend H to a fibrewise map $K: I \times I \times X \to X$ such that $K_{(s,0)} \simeq_B^A k_s$. Then

$$\theta'\theta = k_0 \simeq_B^A K_{(0,0)} \simeq_B^A K_{(0,1)} \simeq_B^A K_{(1,0)} \simeq_B^A k_1 = \mathrm{id}_X,$$

as required.

The main use of (20.11) is to prove

Proposition (20.12). *Let $u: A \to X$ and $v: A \to Y$ be fibrewise cofibrations, where A, X and Y are fibrewise spaces over B. Let $\phi: X \to Y$ be a fibrewise map such that $\phi u = v$. Suppose that ϕ is a fibrewise homotopy equivalence. Then ϕ is a fibrewise homotopy equivalence under A.*

For let $\psi: Y \to X$ be an inverse of ϕ, up to fibrewise homotopy. Since $\psi v = \psi \phi u \simeq_B u$ there exists a fibrewise map $\psi': Y \to X$ such that $\psi' v = u$ and such that $\psi \simeq_B \psi'$. Since $\psi' \phi u = u$ and since $\psi' \phi \simeq_B \mathrm{id}_X$ there exists, by (20.11), a fibrewise map $\psi'': X \to Y$ such that $\psi'' u = v$ and such that $\psi'' \psi' \phi \simeq_B^A \mathrm{id}_X$. Thus ϕ admits a left inverse $\phi' = \psi'' \psi'$, up to fibrewise homotopy under A.

Now ϕ' is a fibrewise homotopy equivalence, since ϕ is a fibrewise homotopy equivalence, and so the same argument, applied to ϕ' instead of ϕ, shows that ϕ' admits a left inverse ϕ'' up to fibrewise homotopy under A. Thus ϕ' admits both a right inverse ϕ and a left inverse ϕ'' up to fibrewise homotopy under A. Hence ϕ' is a fibrewise homotopy equivalence under A, and so ϕ itself is a fibrewise homotopy equivalence under A, as asserted.

Corollary (20.13). *Let* $u: A \to X$ *be a fibrewise cofibration, where* A *and* X *are fibrewise spaces over* B. *If* u *is a fibrewise homotopy equivalence then* u *is a fibrewise homotopy equivalence under* A.

Here we regard A as a fibrewise space under A using the identity, and X as a fibrewise space under A using u.

Proposition (20.14). *Let* $u: A \to X$ *be a fibrewise homotopy equivalence, where* A *and* X *are fibrewise spaces over* B. *Then* A *is a fibrewise deformation retract of the fibrewise mapping cylinder* $M = M_B(u)$.

To see this recall that $u = \rho\sigma_0$, where

$$A \xrightarrow{\sigma_0} M \xrightarrow{\rho} X$$

are as before. Since u and ρ are fibrewise homotopy equivalences, so is σ_0. Also id_A and σ_0 are fibrewise cofibrations, and so σ_0 is a fibrewise homotopy equivalence under A, by (20.12). This proves (20.14).

We already know that any fibrewise map $u: A \to X$ can be factored into the composition of the closed fibrewise embedding σ_0 and the fibrewise homotopy equivalence ρ. We now go on to show that σ_0 is a fibrewise cofibration, so that any fibrewise map can be expressed as the composition of a fibrewise closed cofibration and a fibrewise homotopy equivalence.

To see this consider the retraction

$$r: I \times I \to (\{0\} \times I) \cup (I \times \{0\}),$$

given by projection from the point $(2, 1)$ (see Figure 5). A fibrewise map

$$R: I \times (X + I \times A) \to (\{0\} \times M) \cup (I \times \{0\} \times A)$$

is given by

$$R(t, x) = (0, x) \qquad (x \in X)$$
$$R(s, t, a) = (r(s, t), a) \quad (s, t \in I, a \in A).$$

Since $r(s, 0) = (0, 0)$ for all s we have

$$R(s, 0, a) = (r(s, 0), a) = (0, 0, a) = (0, u(a)) = R(0, u(a)).$$

Thus R induces a fibrewise retraction

$$I \times M \to (\{0\} \times M) \cup (I \times \{0\} \times A),$$

and now our assertion follows from (20.3).

We now come to the product theorem, which has a number of interesting consequences. The proof involves the notion of fibrewise Strøm structure which gives further insight into the nature of the fibrewise cofibration condition and is useful for other purposes also.

Definition (20.15). *A fibrewise Strøm structure on the closed fibrewise pair* (X, A) *over B is a pair* (α, h) *consisting of a map* $\alpha\colon X \to I$ *which is zero throughout A and a fibrewise homotopy* $h\colon I \times X \to X$ *rel A of* id_X *such that* $h(t, x) \in A$ *whenever* $t > \alpha(x)$.

Clearly the existence of a fibrewise Strøm structure is natural, in the sense of §11. Since α can always be replaced by $\alpha' = \min(2\alpha, 1)$ we may substitute $t \geqslant \alpha'(x)$ for $t > \alpha(x)$.

Proposition (20.16). *Let* (X, A) *be a closed fibrewise pair over B. Then* (X, A) *is fibrewise cofibred if and only if* (X, A) *admits a fibrewise Strøm structure.*

For let (α, h) be a fibrewise Strøm structure on (X, A). Then a fibrewise retraction

$$r\colon I \times X \to (\{0\} \times X) \cup (I \times A)$$

is given by

$$r(t, x) = \begin{cases} (0, h(t, x)) & (t \leqslant \alpha(x)) \\ (t - \alpha(x), h(t, x)) & (t \geqslant \alpha(x)), \end{cases}$$

and so (X, A) is fibrewise cofibred, by (20.5).

Conversely, suppose that (X, A) is fibrewise cofibred, and so there exists a fibrewise retraction r as above. Consider the projections

$$I \xleftarrow{r_1} I \times X \xrightarrow{r_2} X.$$

Figure 5

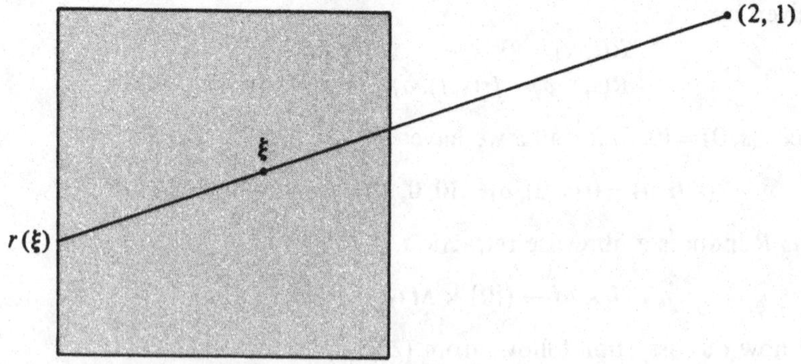

Since I is compact a map $\alpha: X \to I$ is given by

$$\alpha(x) = \sup_{t \in I} |r_1(t, x) - t| \quad (x \in X),$$

and then (α, r_2) constitutes a fibrewise Strøm structure on (X, A).

For example (see §26) if E is an orthogonal $(n-1)$-sphere bundle over B then the pair $(\Gamma_B(E), E)$ is fibrewise cofibred, where $\Gamma_B(E)$ is the associated n-ball bundle.

Proposition (20.17). *Let (X, X') and (Y, Y') be closed fibrewise cofibred pairs over B. Then the closed fibrewise pair*

$$(X, X') \times_B (Y, Y') = (X \times_B Y, X' \times_B Y \cup X \times_B Y')$$

is also fibrewise cofibred.

For let (α, h) and (β, k) be fibrewise Strøm structures on (X, X') and (Y, Y'), respectively. Define $\gamma: X \times_B Y \to I$ by

$$\gamma(x, y) = \min(\alpha(x), \beta(y)) \quad (b \in B, x \in X_b, y \in Y_b)$$

and define $l: I \times X \times_B Y \to X \times_B Y$ by

$$l(t, x, y) = (h(\min(t, \beta(y)), x), k(\min(t, \alpha(x)), y)),$$

where $t \in I$, $b \in B$, $x \in X_b$, $y \in Y_b$. Then (γ, l) constitutes a fibrewise Strøm structure for the closed fibrewise pair $(X, X') \times_B (Y, Y')$, as required.

There is in fact a refinement of these results which is sometimes useful. Let us describe a fibrewise Strøm structure (α, h) on the closed fibrewise pair (X, A) as *strict* if $\alpha < 1$ throughout X. When this condition is satisfied then $h_1 X = A$ and so h_t is a fibrewise deformation retraction. Conversely suppose that A is a fibrewise deformation retract of X and that (X, A) is fibrewise cofibred. Let $H: I \times X \to X$ be a fibrewise deformation retraction and let (α, h) be a fibrewise Strøm structure. Then (α', h') is a strict fibrewise Strøm structure, where

$$\alpha'(x) = \min(\alpha(x), \tfrac{1}{2}) \qquad (x \in X),$$
$$h'(t, x) = H(\min(2t, 1), h(t, x)) \quad (t \in I, x \in X).$$

Returning to the proof of (20.17) we observe that if $\beta < 1$ throughout Y then $\gamma < 1$ throughout $X \times_B Y$, and so we obtain

Proposition (20.18). *Let (X, X') and (Y, Y') be closed fibrewise cofibred pairs over B. If Y' is a fibrewise deformation retract of Y then $X' \times_B Y \cup X \times_B Y'$ is a fibrewise deformation retract of $X \times_B Y$.*

Let (T, T_0) be a closed fibrewise cofibred pair over B. As before, let Φ_B be the endofunctor of the category of fibrewise spaces over B which assigns to each fibrewise space X the fibrewise push-out of the cotriad

$$T \times_B X \leftarrow T_0 \times_B X \rightarrow T_0$$

and similarly with fibrewise maps. Recall that Φ_B transforms each closed fibrewise pair $(X. A)$ into the closed fibrewise pair $(\Phi_B(X), \Phi_B(A))$. If (X, A) is fibrewise cofibred then so is

$$(T, T_0) \times_B (X, A) = (T \times_B X, T_0 \times_B X \cup T \times_B A),$$

by (20.17), and hence so is $(\Phi_B(X), \Phi_B(A))$, by (20.2). Thus we obtain

Proposition (20.19). *If (X, A) is a closed fibrewise cofibred pair over B then so is $(\Phi_B(X), \Phi_B(A))$.*

Corollary (20.20). *If (X, A) is a closed fibrewise cofibred pair over B then so are $(\Gamma_B(X), \Gamma_B(A))$ and $(\Sigma_B(X), \Sigma_B(A))$.*

21. Fibrewise pointed cofibrations

Up to a certain stage the fibrewise pointed version of the theory of the previous chapter is so similar to that of the fibrewise non-pointed theory that it almost seems unnecessary to write it all out. However, the theories diverge quite markedly after that stage and so it seems safest to give full details from the start.

Definition (21.1). *The fibrewise pointed map $u: A \rightarrow X$, where A and X are fibrewise pointed spaces over B, is a fibrewise pointed cofibration if u has the following fibrewise pointed homotopy extension property. Let $f: X \rightarrow E$ be a fibrewise pointed map, where E is a fibrewise pointed space, and let $g: A \rightarrow P_B(E)$ be a fibrewise pointed homotopy such that $\rho_0 g = fu$. Then there exists a fibrewise pointed homotopy $h: X \rightarrow P_B(E)$ such that $\rho_0 h = f$ and $hu = g$.*

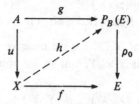

Instead of saying that u is a fibrewise pointed cofibration we may occasionally say that X is a *fibrewise pointed cofibre space* under A: for

example the fibrewise pointed coproduct of fibrewise pointed cofibre spaces under A is a fibrewise pointed cofibre space under A.

An important special case is when A is a subspace of X. In that case we describe (X, A) as a fibrewise pointed cofibred pair when the inclusion $A \to X$ is a fibrewise pointed cofibration. For example (X, B) is automatically a fibrewise pointed cofibred pair, for any fibrewise pointed space X.

Proposition (21.2). *Let $u: A \to X$ be a fibrewise pointed cofibration, where A and X are fibrewise pointed spaces over B. Then the push-out $v: A' \to \xi_* X$ is a fibrewise pointed cofibration for each fibrewise pointed space A' and fibrewise pointed map $\xi: A \to A'$.*

The proof is entirely formal and will be left to serve as an exercise.

In the fibrewise pointed theory an important role is played by the reduced fibrewise mapping cylinder. Let $u: A \to X$ be a fibrewise pointed map, where A and X are fibrewise pointed spaces over B. Recall from §19 that the reduced fibrewise mapping cylinder $\tilde{M} = M_B^B(u)$ is obtained from the ordinary fibrewise mapping cylinder $M = M_B(u)$ by fibrewise collapsing $I \times B$. In the case where A is a subspace of X, with u the inclusion, we have a continuous fibrewise pointed bijection $\tilde{M} \to (\{0\} \times X) \cup (I \tilde{\times} A)$. When A is closed in X the correspondence is topological and so we may identify $\tilde{M}$ with $(\{0\} \times X) \cup (I \tilde{\times} A)$.

For each fibrewise pointed map $u: A \to X$ the reduced fibrewise mapping cylinder $\tilde{M}$ may be interpreted as the reduced fibrewise cone on A, regarded as a fibrewise pointed space over X through u. From the present point of view the fibrewise pointed contractibility of the reduced fibrewise cone is equivalent to the statement that $j: X \to \tilde{M}$ is a fibrewise pointed homotopy equivalence over X.

Consider now the fibrewise pointed map $k: \tilde{M} \to I \tilde{\times} X$ which is derived from the diagram shown below.

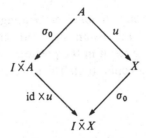

Proposition (21.3). *Let* $u: A \to X$ *be a fibrewise pointed map where A and X are fibrewise pointed spaces over B. Then u is a fibrewise pointed cofibration if and only if the fibrewise pointed map k admits a left inverse* $l: I \tilde{\times} X \to \tilde{M}$.

The proof is similar to that of the corresponding result (20.3) in the previous section and will therefore be omitted. It follows, incidentally, that fibrewise pointed cofibrations are injective.

Corollary (21.4). *Let* (X, A) *be a fibrewise pointed closed pair over B. Then* (X, A) *is fibrewise pointed cofibred if and only if* $(\{0\} \times X) \cup (I \tilde{\times} A)$ *is a fibrewise retract of* $I \tilde{\times} X$.

This in turn implies

Corollary (21.5). *Let* (X, A) *be a fibrewise pointed closed pair over B. If* (X, A) *is fibrewise pointed cofibred over B then* (λ^*X, λ^*A) *is fibrewise pointed cofibred over B' for each space B' and map* $\lambda: B' \to B$.

Proposition (21.6). *Let* (X, A) *be a fibrewise pointed cofibred closed pair over B. Suppose that A is fibrewise pointed contractible. Then the natural projection* $X \to X/_B A$ *to the fibrewise collapse is a fibrewise pointed homotopy equivalence.*

As we have seen in §19 any fibrewise pointed map $u: A \to X$ can be factored into the composition

We already know that any fibrewise pointed map $u: A \to X$ can be factored into the composition

$$A \xrightarrow{\sigma_0} \tilde{M} \xrightarrow{\rho} X.$$

where σ_0 is a fibrewise pointed closed embedding and ρ is a fibrewise pointed homotopy equivalence. By an argument very similar to that used to prove the corresponding result in the previous section we find that σ_0 here is a fibrewise pointed cofibration. The same is true of the following sequence of results.

Proposition (21.7). *Let* $u: A \to X$ *and* $v: A \to Y$ *be fibrewise pointed maps, where A, X and Y are fibrewise pointed spaces over B. Let* $\phi: X \to Y$ *be* *a fibrewise pointed map such that* $\phi u \simeq^A_B v$. *If* u *is a fibrewise pointed cofibration then* $\phi \simeq^A_B \psi$ *for some fibrewise pointed map* $\psi \colon X \to Y$ *such that* $\psi u = v$.

Corollary (21.8). *Let* $u \colon A \to X$ *be a fibrewise pointed cofibration, where* A *and* X *are fibrewise pointed spaces over* B. *If* u *admits a left inverse up to fibrewise pointed homotopy then* u *admits a left inverse.*

Proposition (21.9). *Let* $u \colon A \to X$ *be a fibrewise pointed cofibration, where* A *and* X *are fibrewise pointed spaces over* B. *Let* $\theta \colon X \to X$ *be a fibrewise pointed map such that* $\theta u = u$ *and such that* $\theta \simeq^B_B \mathrm{id}_X$. *Then there exists a fibrewise pointed map* $\theta' \colon X \to X$ *such that* $\theta' u = u$ *and such that* $\theta' \theta \simeq^A_B \mathrm{id}_X$.

Proposition (21.10). *Let* $u \colon A \to X$ *and* $v \colon A \to Y$ *be fibrewise pointed cofibrations, where* A, X *and* Y *are fibrewise pointed spaces over* B. *Let* $\phi \colon X \to Y$ *be a fibrewise pointed map such that* $\phi u = v$. *Suppose that* ϕ *is a fibrewise pointed homotopy equivalence. Then* ϕ *is a fibrewise pointed homotopy equivalence under* A.

Corollary (21.11). *Let* $u \colon A \to X$ *be a fibrewise pointed cofibration, where* A *and* X *are fibrewise pointed spaces over* B. *If* u *is a fibrewise pointed homotopy equivalence then* u *is a fibrewise pointed homotopy equivalence under* A.

Proposition (21.12). *Let* $u \colon A \to X$ *be a fibrewise pointed homotopy equivalence, where* A *and* X *are fibrewise pointed spaces over* B. *Then* A *is a fibrewise pointed deformation retract of the reduced fibrewise mapping cylinder* $M^B_B(u)$.

The remainder of this section is devoted to an outline of the fibrewise version of a theory due to Puppe [54]. We shall be concerned with sequences

$$X_1 \xrightarrow{f_1} X_2 \xrightarrow{f_2} X_3 \xrightarrow{f_3} \cdots$$

of fibrewise pointed spaces and fibrewise pointed maps, over B. Let us describe such a sequence as *exact* if the induced sequence

$$\pi_B^B(X_1, E) \leftarrow \underset{f_2^*}{\pi_B^B(X_2, E)} \leftarrow \underset{f_2^*}{\pi_B^B(X_3, E)} \leftarrow \cdots$$

of pointed sets is exact, for all fibrewise pointed spaces E.

If $\phi: 2 \to Y$ is a fibrewise pointed map, where X and Y are fibrewise pointed spaces over B, then the *reduced fibrewise mapping cone*, or *fibrewise pointed homotopy cofibre*, of ϕ, is defined to be the push-out $\Gamma_B^B(\phi)$ of the cotriad

$$\Gamma_B^B(X) \underset{\sigma_1}{\leftarrow} X \underset{\phi}{\to} Y.$$

This is equivalent to the fibrewise collapse $M_B^B(\phi)/_B X$. Now $\Gamma_B^B(\phi)$ comes equipped with a fibrewise embedding

$$\phi': Y \to \Gamma_B^B(\phi),$$

and we have

Proposition (21.13). *The sequence*

$$\pi_B^B(X, E) \underset{\phi^*}{\leftarrow} \pi_B^B(Y, E) \underset{\phi'^*}{\leftarrow} \pi_B^B(\Gamma_B^B(\phi), E)$$

of pointed sets is exact, for all fibrewise pointed spaces E.

In fact $\phi' \circ \phi$ extends to a fibrewise pointed map $\Gamma_B^B(X) \to \Gamma_B^B(\phi)$, in an obvious way, and so is fibrewise pointed nulhomotopic. This shows that $\phi^* \circ \phi'^* = 0$. Conversely suppose that $f: X \to E$ is a fibrewise pointed map such that $\phi \circ f$ is fibrewise pointed nulhomotopic. Let $h: I \overset{\sim}{\times} X \to E$ be a fibrewise pointed nulhomotopy of $\phi \circ f$. Then h and ϕ together form a fibrewise pointed map $g: \Gamma_B^B(\phi) \to E$ such that $g \circ \phi' = f$. This completes the proof.

Proposition (21.14). *Let $u: A \to X$ be a fibrewise pointed cofibration, where A and X are fibrewise pointed spaces over B. Then the natural projection*

$$\Gamma_B^B(u) \to \Gamma_B^B(u)/_B \Gamma_B^B(A) = X/_B A$$

is a fibrewise pointed homotopy equivalence.

We begin by showing that there exists a fibrewise pointed homotopy of the identity on $\Gamma_B^B(u)$ which deforms $\Gamma_B^B(A)$ over itself into the section. For consider the fibrewise pointed nulhomotopy $h_t: \Gamma_B^B(A) \to \Gamma_B^B(u)$ of the inclusion $\Gamma_B^B(A) \to \Gamma_B^B(u)$, given by

$$h_t(s, x) = (s(1-t), x) \quad ((s, x) \in \Gamma_B^B(A), t \in I).$$

Since u is a fibrewise pointed cofibration we can extend $h_t u$ to a fibrewise pointed homotopy $g_t \colon X \to \Gamma^B_B(\phi)$ of the inclusion. Together h_t and g_t form a fibrewise pointed homotopy of the identity on $\Gamma^B_B(u)$ which deforms $\Gamma^B_B(A)$ over itself into the section as required. Now the same argument as was used to prove (21.6) completes the proof.

Notice, incidentally, that the fibrewise pointed homotopy equivalence in (21.14) transforms u' into the natural projection $X \to X /_B A$. Returning to the general case, where $\phi \colon X \to Y$, we now prove

Proposition (21.15). *The embedding $\phi' \colon Y \to \Gamma^B_B(\phi)$ is a fibrewise pointed cofibration.*

In fact the embedding is the fibrewise pointed push-out with respect to $\phi \colon X \to Y$ of the embedding $X \to \Gamma^B_B(X)$, which is a fibrewise pointed cofibration since $\Gamma^B_B(X) = M^B_B(p)$ where $p \colon X \to B$ is the projection.

By combining the last two results we see that the reduced fibrewise pointed mapping cone $\Gamma^B_B(\phi')$ is fibrewise pointed homotopy equivalent to the reduced fibrewise suspension

$$\Gamma^B_B(\phi')/_B \Gamma^B_B(\phi) = \Gamma^B_B(\phi)/_B Y = \Sigma^B_B(X),$$

and in the process $(\phi')'$ is transformed into a fibrewise pointed map $\phi'' \colon \Gamma^B_B(\phi) \to \Sigma^B_B(X)$. Repeating the process we find that $\Gamma^B_B((\phi')')$ is fibrewise pointed homotopy equivalent to the reduced fibrewise suspension $\Sigma^B_B(Y)$, and in the process $((\phi')')'$ is transformed into the reduced fibrewise suspension $\Sigma^B_B(\phi) \colon \Sigma^B_B(X) \to \Sigma^B_B(Y)$ of ϕ composed with the fibrewise reflection $(t, x) \to (1 - t, x)$. This last does not affect exactness so we conclude that the sequence

$$X \to Y \to \Gamma^B_B(\phi) \to \Sigma^B_B(X) \to \Sigma^B_B(Y) \to \cdots$$

is exact, in the sense we have given it.

It is not difficult to see that the fibrewise pointed homotopy type (in the obvious sense) of this exact sequence depends only on the fibrewise pointed homotopy class of ϕ. In particular, if ϕ is fibrewise pointed nulhomotopic then the sequence has the same fibrewise pointed homotopy type as in the case of the fibrewise constant map

$$X \xrightarrow{c} Y \to Y \vee_B \Sigma^B_B(X) \to \Sigma^B_B(X) \to \cdots.$$

We conclude this section with an application of some of these ideas, the significance of which will become clearer in the next section. Let X and Y be fibrewise pointed spaces over B. Consider the embedding

$$u \colon X \vee_B Y \to X \times_B Y$$

of the fibrewise pointed coproduct in the fibrewise topological product. We denote the reduced fibrewise mapping cone of u by $X \bar{\wedge}_B Y$, so that the first few stages in the fibrewise mapping sequence of u read

$$X \vee_B Y \xrightarrow{u} X \times_B Y \xrightarrow{v} X \bar{\wedge}_B Y \xrightarrow{w} \Sigma_B^B(X) \vee_B \Sigma_B^B(Y),$$

where $v = u'$ and $w = u''$. I assert that w here is fibrewise pointed nulhomotopic. This being so it will follow that the induced function

$$v^* : \pi_B^B(X \bar{\wedge}_B Y, E) \to \pi_B^B(X \times_B Y, E)$$

is injective for all fibrewise pointed spaces E, and hence that $\Gamma_B^B(w)$ has the same fibrewise pointed homotopy type as the fibrewise pointed coproduct

$$\Sigma_B^B(X) \vee_B \Sigma_B^B(Y) \vee_B \Sigma_B^B(X \bar{\wedge}_B Y).$$

The proof of the assertion about w is as follows.

For any fibrewise pointed inclusion $w : A \to Z$, where A and Z are fibrewise pointed spaces over B, we may identify $\Gamma_B^B(w)$ with the subset of $\Gamma_B^B(Z)$ consisting of points (s, z) of $\Gamma_B^B(Z)$ such that either $s \in I$ and $z \in A$ or $s = 1$ and $z \in Z$. Although $\Gamma_B^B(w)$ does not necessarily have the induced topology the inclusion $\Gamma_B^B(w) \to \Gamma_B^B(Z)$ is always continuous.

In this notation the fibrewise pointed map

$$w : X \bar{\wedge}_B Y \to \Sigma_B^B(X) \vee_B \Sigma_B^B(Y),$$

is given by

$$w(s, x) = (s, x), \qquad w(s, y) = (s, y), \qquad w(1, x, y) = b,$$

where $s \in I$, $b \in B$, $x \in X_b$, $y \in Y_b$. Let us write

$$\bar{\Sigma}_B^B(X) = \Sigma_B^B(X)/_B\{(s, x) : x \in X_b, s \leqslant \tfrac{1}{2}\}$$

$$\underline{\Sigma}_B^B(Y) = \Sigma_B^B(Y)/_B\{(s, y) : y \in Y_b, s \geqslant \tfrac{1}{2}\}.$$

Clearly the natural projections

$$\Sigma_B^B(X) \to \bar{\Sigma}_B^B(X), \qquad \Sigma_B^B(Y) \to \underline{\Sigma}_B^B(Y)$$

are fibrewise pointed homotopy equivalences, and hence the natural projection

$$\rho : \Sigma_B^B(X) \vee_B \Sigma_B^B(Y) \to \bar{\Sigma}_B^B(X) \vee_B \underline{\Sigma}_B^B(Y)$$

is a fibrewise pointed homotopy equivalence. So consider the fibrewise pointed map

$$\xi : \Gamma_B^B(X \times_B Y) \to \bar{\Sigma}_B^B(X) \vee_B \underline{\Sigma}_B^B(Y)$$

which is given by

$$\xi(s, x, y) = \begin{cases} (s, y) & (s \leqslant \tfrac{1}{2}, b \in B, y \in Y_b) \\ (s, x) & (s \geqslant \tfrac{1}{2}, b \in B, x \in X_b). \end{cases}$$

The composite of ξ with the fibrewise pointed map

$$X \bar{\wedge}_B Y \rightarrow \Gamma_B^B(X \times_B Y)$$

coincides with ρw. Since $\Gamma_B^B(X \times_B Y)$ is fibrewise pointed conractible, ρw is fibrewise pointed nulhomotopic and so w is fibrewise pointed nulhomotopic, as asserted, since ρ is a fibrewise pointed homotopy equivalence.

Before pursuing these investigations any further we need to make some more definitions and begin a new section for this purpose.

22. Fibrewise non-degenerate spaces

In the fibrewise pointed theory difficulties are encountered if one tries to find analogues for the results towards the end of §20. For example there appears to be no satisfactory concept of fibrewise pointed Strøm structure and results which are proved by that method, such as the fibrewise product theorem and its consequences, seem to be generally false in the fibrewise pointed theory. In seeking a class of well-behaved fibrewise pointed spaces an obvious possibility is to use

Definition (22.1). *The fibrewise pointed space* X *over* B *is fibrewise well-pointed* if the section $B \rightarrow X$ is a fibrewise cofibration.†

For example B is fibrewise well-pointed, as a fibrewise pointed space over itself with the identity as section and projection. For another example let (X, A) be a closed fibrewise cofibred pair over B; then the fibrewise collapse $X/_B A$ is fibrewise well-pointed, by (20.2). Clearly the fibrewise pointed coproduct of fibrewise well-pointed spaces is fibrewise well-pointed. Also it follows from (20.17) that the fibrewise topological product and the fibrewise smash product of fibrewise well-pointed spaces are fibrewise well-pointed. Again (see §26) orthogonal sphere-bundles over B are fibrewise well-pointed.

The condition is also natural in the sense that if X is fibrewise well-pointed over B then the fibrewise pull-back λ^*X is fibrewise well-pointed over B' for each space B' and continuous function $\lambda: B' \rightarrow B$, also the restriction $X_{B'}$ is fibrewise well-pointed over B' for each subspace B' of B.

Let (T, T_0) be a closed cofibred pair over B. As before consider the endofunctor Φ_B of the category of fibrewise spaces over B which assigns

---

† The term *well-sectioned* may also be used instead of fibrewise well-pointed.

to each fibrewise space X the fibrewise push-out of the cotriad

$$T \times_B X \leftarrow T_0 \times_B X \rightarrow T_0.$$

Recall that when X is fibrewise pointed the transform of the section embeds $T = \Phi_B(B)$ in $\Phi_B(X)$ as a closed subspace so that the fibrewise collapse

$$\Phi_B^B(X) = \Phi_B(X)/_B\Phi_B(B)$$

is defined, as a fibrewise pointed space. Suppose that T is fibrewise contractible. Then (20.8) and (20.19) imply that the natural projection

$$\Phi_B(X) \rightarrow \Phi_B^B(X)$$

is a fibrewise homotopy equivalence for all fibrewise well-pointed X. This applies, in particular, to the fibrewise cone functor Γ_B, given by $(T, T_0) = B \times (I, \{0\})$, and to the fibrewise suspension functor Σ_B, given by $(T, T_0) = B \times (I, \dot{I})$.

It turns out however that there is a weaker condition than fibrewise well-pointed, as follows, which suits the applications better. Let X be a fibrewise pointed space over B, with section $s: B \rightarrow X$. We regard the fibrewise mapping cylinder $M_B(s)$ of s as a fibrewise pointed space with section σ_1, and denote it by $\check{X}_B$. Note that the fibrewise mapping cylinder is not reduced, and so the inclusion $\sigma: X \rightarrow \check{X}_B$ is a fibrewise map, not a fibrewise pointed map. In fact σ is a fibrewise homotopy equivalence, as we have seen. Of course the natural projection $\rho: \check{X}_B \rightarrow X$, which fibrewise collapses $M_B(B) = I \times B$, is a fibrewise pointed map and a fibrewise homotopy equivalence.

Definition (22.2). *The fibrewise pointed space X over B is fibrewise non-degenerate if the natural projection $\rho: \check{X}_B \rightarrow X$ is a fibrewise pointed homotopy equivalence.*

Note that $\check{X}_B$ itself is always fibrewise non-degenerate, so that every fibrewise pointed space has the same fibrewise homotopy type as a fibrewise non-degenerate space. This indicates that the class of fibrewise non-degenerate spaces is not too restrictive. Moreover, the class is natural, in the sense of §11.

By a *fibrewise Puppe structure* on a fibrewise pointed space X over B I mean a pair (α, U), where U is a neighbourhood of B in X which is fibrewise contractible in X and $\alpha: X \rightarrow I$ is a map such that $\alpha = 1$ throughout B and $\alpha = 0$ away from U. For example, suppose that (α, h) is a fibrewise Strøm structure on (X, B). Then (α, U) is a fibrewise Puppe structure on X, where $U = \alpha^{-1}(0, 1]$ and the fibrewise contraction is given by the restriction of h. Clearly the existence of a fibrewise Puppe structure is natural, in the sense of §11.

Proposition (22.3). *Let X be a fibrewise pointed space over B. Then X is fibrewise non-degenerate if and only if X admits a fibrewise Puppe structure.*

For suppose that $\rho: \check{X}_B \to X$ is a fibrewise pointed homotopy equivalence, with inverse $\rho': X \to \check{X}_B$ up to fibrewise pointed homotopy. Take U to be the inverse image under ρ' of the open cylinder $(0, 1] \times B \subset M_B(s)$. Then the inclusion $U \to X$ is fibrewise pointed nulhomotopic since $\rho'|U$ is fibrewise pointed nulhomotopic and $\rho\rho'|U$ is fibrewise pointed homotopic to the inclusion. Take $\alpha: X \to I$ to be the composition

$$X \xrightarrow{\rho'} \check{X}_B \to \check{X}_B/_B X = I \times B \xrightarrow{\pi_1} I,$$

where the middle stage is fibrewise collapse. Then $\alpha = 1$ on B and $\alpha = 0$ away from U, so that (α, U) constitutes a fibrewise Puppe structure on X.

Conversely suppose that (α, U) is a fibrewise Puppe structure on X. Without real loss of generality we may suppose that U is a closed neighbourhood of B, since otherwise we can replace (α, U) by the fibrewise Puppe structure (α', U'), where $U' = \alpha^{-1}[0, \frac{1}{2}]$ and $\alpha' = \min(1 - 2\alpha, 1)$. Let $f: I \tilde{\times} U \to X$ be a fibrewise pointed nulhomotopy of the inclusion. Then a fibrewise pointed homotopy $f': I \tilde{\times} U \to \check{X}_B$ is given by

$$f'(t, x) = \begin{cases} f(2t, x) & (t \leqslant \frac{1}{2}, x \in X_b, b \in B) \\ (2t - 1, b) & (t \geqslant \frac{1}{2}, x \in X_b, b \in B). \end{cases}$$

Now a fibrewise pointed homotopy $g_t: \check{X}_B \to \check{X}_B$ of the identity into $\sigma\rho$ is given on $X \subset \check{X}_B$ by

$$g_t(x) = \begin{cases} x & (x \notin U_b, b \in B), \\ f'(t \cdot \alpha(x), x) & (x \in U_b, b \in B), \end{cases}$$

and on $I \times B \subset \check{X}_B$ by

$$g_t(s, b) = \begin{cases} (s, b) & (t \leqslant \frac{1}{2}, s \in I, b \in B) \\ (1 - (1 - s)(2 - 2t), b) & (t \geqslant \frac{1}{2}, s \in I, b \in B). \end{cases}$$

Therefore X is fibrewise non-degenerate and the proof is complete.

In view of (22.3) it is hardly surprising that the behaviour of fibrewise non-degenerate spaces resembles that of fibrewise well-pointed spaces. For example we have

Proposition (22.4). *If X and Y are fibrewise non-degenerate spaces over B then so is the fibrewise topological product $X \times_B Y$.*

For let (α, U) and (β, V) be fibrewise Puppe structures on X and Y respectively. Then (γ, W) is a fibrewise Puppe structure on $X \times_B Y$, where

$W = U \times_B V$ and where $\gamma: X \times_B Y \to I$ is given by

$$\gamma(x, y) = \alpha(x) \cdot \beta(y) \quad (x \in X_b, \, y \in Y_b, \, b \in B).$$

We now return to the situation being considered towards the end of the previous section and prove

Proposition (22.5). *Let X and Y be fibrewise non-degenerate spaces over B. Then the natural projection*

$$X \bar{\wedge}_B Y \to (X \bar{\wedge}_B Y)/_B \Gamma_B^B(X \vee_B Y) = (X \times_B Y)/_B(X \vee_B Y) = X \wedge_B Y$$

is a fibrewise pointed homotopy equivalence.

By combining this with the result established at the end of the previous section we obtain

Corollary (22.6). *If X and Y are fibrewise non-degenerate over B then $\Sigma_B^B(X \times_B Y)$ has the same fibrewise pointed homotopy type as the fibrewise pointed coproduct*

$$\Sigma_B^B(X) \vee_B \Sigma_B^B(Y) \vee_B \Sigma_B^B(X \wedge_B Y).$$

In fact a fibrewise pointed homotopy equivalence is defined by taking the fibrewise track sum, in some order, of the reduced fibrewise suspensions of the projections

$$X \leftarrow X \times_B Y \to Y$$

and the natural projection

$$X \times_B Y \to X \wedge_B Y.$$

This can be checked without difficulty by referring back to the details given in the previous section.

Unfortunately the proof of (22.5), and hence (22.6), is rather lengthy. It depends on

Lemma (22.7). *If X and Y are fibrewise pointed spaces over B then the closed fibrewise pair*

$$(\check{X}_B \times_B \check{Y}_B, \, \check{X}_B \vee_B \check{Y}_B)$$

is fibrewise cofibred.

We defer the proof of (22.7) for a moment in order to show how (22.5) follows from (22.7). In (22.5) X and Y are fibrewise non-degenerate, i.e.

the natural projections

$$\rho_1 : \check{X}_B \to X, \qquad \rho_2 : \check{Y}_B \to Y$$

are fibrewise pointed homotopy equivalences. Consider the diagram shown below where r and r' are the fibrewise collapses, as in (22.5), and where $q = \rho_1 \wedge \rho_2$ and $\bar{q} = \rho_1 \bar{\wedge} \rho_2$.

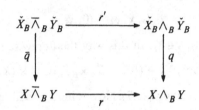

Since ρ_1 and ρ_2 are fibrewise pointed homotopy equivalences, so are q and $\bar{q}$. Also r' is a fibrewise pointed homotopy equivalence, by (21.14) and (22.7). Since the diagram is commutative we conclude that r is a fibrewise pointed homotopy equivalence, as asserted.

It only remains, then, to establish the lemma, which we do by defining a fibrewise retraction

$$R : I \times \check{X}_B \times_B \check{Y}_B \to (\{0\} \times \check{X}_B \times_B \check{Y}_B) \cup (I \times (\check{X}_B \vee_B \check{Y}_B)).$$

Recall that $\check{X}_B$ and $\check{Y}_B$ are fibrewise quotient spaces of $I \times B +_B X$ and $I \times B +_B Y$, respectively. Since the natural projections are proper it follows at once that their fibrewise product

$$(I \times B +_B X) \times_B (I \times B +_B Y) \to \check{X}_B \times_B \check{Y}_B$$

is also a fibrewise quotient map. We may therefore define R by defining a fibrewise map on the four summands of the domain, respecting the identifications, and then taking the induced fibrewise map of the codomain. So let $\phi_1, \phi_2 : I \times I \to I$ be the components of the central projection

$$\phi : I \times I \to (I \times \{0\}) \cup (\{1\} \times I)$$

from the point $(0, 2)$, and let

$$\psi_1, \psi_2, \psi_3 : I \times I \times I \to I$$

be the components of the central projection

$$\psi : I \times I \times I \to (I \times I \times \{0\}) \cup (I \times \{1\} \times I) \cup (\{1\} \times I \times I)$$

from the point $(0, 0, 2)$. Then a fibrewise map of the four summands with the required properties is given by

$$(s, (t, b), (u, b)) \mapsto (\psi_1(s, t, u), (\psi_2(s, t, u), b), (\psi_3(s, t, u), b)),$$

for $s, t, u \in I$, $b \in B$; by

$$(s, (t, b), y) \mapsto ((\phi_1(s, t), b), (\phi_2(s, t), y))$$

for $s, t \in I$, $b \in B$, $y \in Y_b$; by

$$(s, x, (t, b)) \mapsto ((\phi_1(s, t), x), (\phi_2(s, t), b))$$

for $s, t \in I$, $b \in B$, $x \in X_b$; by

$$(s, x, y) \mapsto (0, x, y)$$

for $s \in I$, $b \in B$, $x \in X_b$, $y \in Y_b$. In this way the fibrewise retraction

$$R: I \times \check{X}_B \times_B \check{Y}_B \to (\{0\} \times \check{X}_B \times_B \check{Y}_B) \cup (I \times (\check{X}_B \vee_B \check{Y}_B))$$

is defined, and so we obtain (22.7). Now R induces a fibrewise retraction

$$\bar{R}: I \times (\check{X}_B \wedge_B \check{Y}_B) \to I \times B \vee_B (\check{X}_B \wedge_B \check{Y}_B)$$

and so a fibrewise pointed homotopy

$$h_t : (\check{X}_B \wedge_B \check{Y}_B)_B^{\vee} \to (\check{X}_B \wedge \check{Y}_B)_B^{\vee}$$

of the identity is given by

$$h_t(x, y) = \bar{R}(t, x, y) \qquad (t \in I, b \in B, x \in X_b, y \in Y_b)$$
$$h_t(s, b) = (1 - (1 - s)(1 - t), b) \quad (s, t \in I, b \in B).$$

Since h_t fibrewise contracts the cylinder $I \times B$ over itself into $\{0\} \times B$ we conclude that the natural projection

$$(\check{X}_B \wedge_B \check{Y}_B)_B^{\vee} \to \check{X}_B \wedge_B \check{Y}_B$$

is a fibrewise pointed homotopy equivalence, i.e. that $\check{X}_B \wedge_B \check{Y}_B$ is fibrewise non-degenerate. Moreover, $\check{X}_B \wedge_B \check{Y}_B$ and $X \wedge_B Y$ have the same fibrewise pointed homotopy type, as we have seen, and so we obtain

Proposition (22.8). *If X and Y are fibrewise non-degenerate spaces over B then so is the fibrewise smash product $X \wedge_B Y$.*

With the help of (22.8) we can prove that the fibrewise smash product is associative, up to fibrewise pointed homotopy equivalence, for the class of fibrewise non-degenerate spaces, as follows.

Proposition (22.9). *Let X, Y and Z be fibrewise non-degenerate spaces over B. Then $(X \wedge_B Y) \wedge_B Z$ and $X \wedge_B (Y \wedge_B Z)$ have the same fibrewise pointed homotopy type.*

To prove (22.9) we consider the subspace

$$X \Delta_B Y \Delta_B Z = B \times_B Y \times_B Z \cup X \times_B B \times_B Z \cup X \times_B Y \times_B B$$

of $X \times_B Y \times_B Z$ (the fibrewise 'fat wedge') and form the fibrewise quotient

$$X \wedge_B Y \wedge_B Z = (X \times_B Y \times_B Z)/_B(X \Delta_B Y \Delta_B Z)$$

by fibrewise collapsing the subspace. As a fibrewise pointed set $X \wedge_B Y \wedge_B Z$ coincides with $(X \wedge_B Y) \wedge_B Z$ and $X \wedge_B (Y \wedge_B Z)$ but in general the fibrewise topologies are different. There are, however, fibrewise pointed maps

$$(X \wedge_B Y) \wedge_B Z \leftarrow X \wedge_B Y \wedge_B Z \rightarrow X \wedge_B (Y \wedge_B Z),$$

induced by the identity. I assert that these are fibrewise pointed homotopy equivalences, from which (22.9) will follow at once. To prove the assertion it is sufficient to establish

Proposition (22.10). *Let X, Y and Z be fibrewise pointed spaces over B. Then the fibrewise pointed map*

$$\zeta \colon \check{X}_B \wedge_B \check{Y}_B \wedge_B Z \rightarrow (\check{X}_B \wedge_B \check{Y}_B) \wedge_B Z$$

induced by the identity is a fibrewise pointed homotopy equivalence.

First consider the fibrewise pointed map $u \colon \check{X}_B \rightarrow \check{X}_B$ which leaves X fixed and on $I \times B$ is given by

$$u(t, b) = (\max(0, 2t - 1), b) \quad (t \in I, b \in B).$$

If $v \colon \check{Y}_B \rightarrow \check{Y}_B$ is defined similarly then a fibrewise pointed function

$$\eta \colon (\check{X}_B \wedge_B \check{Y}_B) \wedge_B \check{Z}_B \rightarrow \check{X}_B \wedge_B \check{Y}_B \wedge_B \check{Z}_B$$

is induced by $u \times v \times$ id. I assert that η is continuous.

For consider the closed subsets A_1, A_2 of $\check{X}_B \wedge_B \check{Y}_B$ given by

$$A_1 = \{((t, b), y) \quad (b \in B, y \in Y_b, t \geqslant \tfrac{1}{2})$$

$$(x, (t, b)) \quad (b \in B, x \in X_b, t \geqslant \tfrac{1}{2})\}$$

$$A_2 = \{(x, y) \quad (b \in B, x \in X_b, y \in Y_b)$$

$$(x, (t, b)) \quad (b \in B, y \in Y_b, t \leqslant \tfrac{1}{2})$$

$$((t, b), y) \quad (b \in B, x \in X_b, t \leqslant \tfrac{1}{2})$$

$$((t, b), (t, b)) \quad (b \in B, t \leqslant \tfrac{1}{2})\}$$

Since $(\check{X}_B \wedge_B \check{Y}_B) \vee_B \check{Z}_B$ is closed in $(\check{X}_B \wedge_B \check{Y}_B) \times_B \check{Z}_B$ the natural projection

$$(\check{X}_B \wedge_B \check{Y}_B) \times_B \check{Z}_B \to (\check{X}_B \wedge_B \check{Y}_B) \wedge_B \check{Z}_B$$

is closed. Thus the images M_i ($i = 1, 2$) of $A_i \times_B \check{Z}_B$ under the projection are closed. Now η is fibrewise constant on M_1, and therefore continuous. Moreover M_2 can be identified with the fibrewise quotient

$$A_2 \times_B \check{Z}/_B A_2 \times_B (B \times 1),$$

which has the same relative topology whether regarded as a subspace of $(\check{X}_B \wedge_B \check{Y}_B) \times_B \check{Z}_B$ or as a subspace of the fibrewise quotient space of $\check{X}_B \times_B \check{Y}_B \times_B \check{Z}_B$. Since η is continuous on M_2 regarded as a subspace of the latter we obtain that η is continuous on M_2 regarded as a subspace of the former. Since M_1 and M_2 cover $(\check{X}_B \wedge_B \check{Y}_B) \wedge \check{Z}_B$ we conclude that η is continuous, as asserted.

Note that $\zeta\eta$ and $\eta\zeta$ have the same underlying set function. I assert that $\zeta\eta$ is fibrewise pointed homotopic to the identity on $(\check{X}_B \wedge_B \check{Y}_B) \wedge_B \check{Z}_B$, while $\eta\zeta$ is fibrewise pointed homotopic to the identity on $\check{X}_B \wedge_B \check{Y}_B \wedge_B \check{Z}_B$. To see this, consider the fibrewise pointed homotopy $u_s \colon \check{X}_B \to \check{X}_B$ which leaves X fixed and on $I \times B$ is given by

$$u_s(t, b) = (\max(0, (2t - s)/(2 - s)), b) \quad (s, t \in I, b \in B).$$

If $v_s \colon \check{Y}_B \to \check{Y}_B$ is defined similarly then $u_s \times v_s$ defines a fibrewise pointed homotopy $H_s \colon \check{X}_B \times_B \check{Y}_B \to \check{X}_B \times_B \check{Y}_B$. Now on the one hand $H_s \times \mathrm{id}_Z$ induces a fibrewise pointed homotopy

$$(\check{X}_B \wedge_B \check{Y}_B) \wedge_B \check{Z}_B \to (\check{X}_B \wedge_B \check{Y}_B) \wedge_B \check{Z}_B$$

of the identity into $\eta\zeta$, while on the other hand $H_s \times \mathrm{id}$ induces a fibrewise pointed homotopy

$$\check{X}_B \wedge_B \check{Y}_B \wedge_B \check{Z}_B \to \check{X}_B \wedge_B \check{Y}_B \wedge_B \check{Z}_B$$

of the identity into $\zeta\eta$. This proves (22.10), and hence (22.9).

Although quite a number of steps are involved in establishing the existence of the fibrewise pointed homotopy equivalence in (22.9), each of these steps involves showing that a specific fibrewise pointed map is a fibrewise pointed homotopy equivalence. It follows that the end result is unique, up to fibrewise pointed homotopy. It also follows that the construction is functorial, up to fibrewise pointed homotopy. Specifically, let

$$\theta \colon X \to X', \qquad \phi \colon Y \to Y', \qquad \psi \colon Z \to Z'$$

be fibrewise pointed maps, where X, Y, Z and X', Y', Z' are fibrewise non-degenerate spaces. Then the diagram shown below is commutative up to fibrewise pointed homotopy, where the horizontals are the fibrewise pointed homotopy equivalences arising from the proof of (22.9)

$$
\begin{array}{ccc}
X \wedge_B (Y \wedge_B Z) & \longrightarrow & (X \wedge_B Y) \wedge_B Z \\
\theta \wedge (\phi \wedge \psi) \Big\downarrow & & \Big\downarrow (\theta \wedge \phi) \wedge \psi \\
X' \wedge_B (Y' \wedge_B Z') & \longrightarrow & (X' \wedge_B Y') \wedge_B Z'
\end{array}
$$

Note that (22.6) can obviously be iterated, so as to yield

Proposition (22.11). *Let* $X_1, \ldots, X_n$ *be fibrewise non-degenerate spaces over B. Then* $\Sigma_B^B(X_1 \times_B \cdots \times_B X_n)$ *has the same fibrewise pointed homotopy type as the fibrewise pointed coproduct*

$$
\bigvee_N {}_B \Sigma_B^B \bigwedge_{i \in N} {}_B X_i,
$$

where N runs through all non-empty subsets of the integers 1 to n.

In fact a fibrewise pointed homotopy equivalence is given by taking the fibrewise track sum of the reduced fibrewise suspensions of the natural projections

$$
X_1 \times_B \cdots \times_B X_n \to \bigwedge_{i \in N} {}_B X_i,
$$

where $X_i \to B$ by the projection when $i \notin N$, and X_i is mapped by the identity when $i \in N$.

By taking the reduced fibrewise suspension of an appropriate one of these projections and precomposing we obtain a monomorphism

$$
\pi_B^B(\Sigma_B^B(\bigwedge_{i \in N} {}_B X_i), E) \to \pi_B^B(\Sigma_B^B(X_1 \times_B \cdots \times_B X_n), E),
$$

with image a normal subgroup. In what is to follow it is often convenient to regard each of the groups $\pi_B^B(\Sigma_B^B(\bigwedge_{i \in N} {}_B X_i), E)$ as being embedded in $\pi_B^B(\Sigma_B^B(X_1 \times_B \cdots \times_B X_n), E)$ as a normal subgroup. Note that if the fibrewise cogroup-like structure on $\Sigma_B^B(X_i)$ is fibrewise homotopy-commutative for some $i \in N$ then the subgroup $\pi_B^B(\Sigma_B^B(\bigwedge_{i \in N} {}_B X_i), E)$ is commutative, since $\Sigma_B^B(\bigwedge_{i \in N} {}_B X_i)$ has the same fibrewise pointed homotopy type as

$$
\Sigma_B^B(X_i) \wedge_B \bigwedge_{j \in N, j \neq i} {}_B X_j,
$$

as we have seen, and the fibrewise cogroup-like structure is preserved.

By way of application we give an outline of the theory of the fibrewise Whitehead product, beginning with some remarks of a general nature. Consider a family $\{X_i\}$ $(i = 1, \ldots, n)$ of fibrewise non-degenerate spaces. Each projection

$$
\pi_i \colon \prod_B X_i \to X \quad (i = 1, \ldots, n)
$$

has a right inverse. Hence the reduced fibrewise suspension of π_i has a right inverse and therefore embeds

$$\pi_B^B(\Sigma_B^B(X_i), E) \quad (i = 1, \ldots, n)$$

as a normal subgroup of the group

$$\pi_B^B(\Sigma_B^B(\textstyle\prod_B X_i), E),$$

for each fibrewise pointed space E. Other significant normal subgroups can be identified as follows. First take the case $n = 2$; in that case the reduced fibrewise suspension of the projection

$$X_1 \times_B X_2 \to X_1 \wedge_B X_2$$

is one of the fibrewise pointed maps involved in the splitting of $\Sigma_B^B(X_1 \times_B X_2)$, as in (22.6). Therefore the image of

$$\pi_B^B(\Sigma_B^B(X_1 \wedge_B X_2), E)$$

under the corresponding induced homomorphism is a normal subgroup of the group $\pi_B^B(\Sigma_B^B(X_1 \times_B X_2), E)$. Next take the case $n = 3$; in this case the reduced fibrewise suspension of the projection

$$X_1 \times_B X_2 \times_B X_3 \to X_1 \wedge_B X_2 \wedge_B X_3$$

is one of the fibrewise pointed maps involved in the splitting of $\Sigma_B^B(X_1 \times_B X_2 \times_B X_3)$, as in (22.11). Therefore the image of

$$\pi_B^B(\Sigma_B^B(X_1 \wedge_B X_2 \wedge_B X_3), E)$$

under the corresponding induced homomorphism is a normal subgroup of the group $\pi_B^B(\Sigma_B^B(X_1 \times_B X_2 \times_B X_3), E)$. Of course, the group also contains normal subgroups of the type described in the case $n = 2$, such as

$$\pi_B^B(\Sigma_B^B(X_1 \wedge_B (X_2 \times_B X_3)), E).$$

The corresponding statements in the general case may be formulated but this is as far as we need to go for present purposes.

Now let $\alpha_i \in \pi_B^B(\Sigma_B^B(X_i), E)$ $(i = 1, 2)$. We form the commutator $[\alpha_1, \alpha_2]$ in $\pi_B^B(\Sigma_B^B(X_1 \times_B X_2), E)$. The commutator vanishes under the homomorphism

$$\pi_B^B(\Sigma_B^B(X_1 \times_B X_2), E) \to \pi_B^B(\Sigma_B^B(X_1 \vee_B X_2), E)$$

$$= \pi_B^B(\Sigma_B^B(X_1), E) \times \pi_B^B(\Sigma_B^B(X_2), E),$$

induced by the reduced fibrewise suspension of the inclusion

$$X_1 \vee_B X_2 \to X_1 \times_B X_2.$$

The commutator therefore lies in the kernel of this homomorphism, i.e. the normal subgroup

$$\pi_B^B(\Sigma_B^B(X_1 \wedge_B X_2), E).$$

Without changing notation we refer to the element

$$[\alpha_1, \alpha_2] \in \pi_B^B(\Sigma_B^B(X_1 \wedge_B X_2), E)$$

as the *fibrewise Whitehead product*† of α_1 and α_2. We have at once that

(22.12) $$[\alpha_2, \alpha_1] = -(\Sigma_B^B t)^*[\alpha_1, \alpha_2],$$

where $t: X_1 \wedge_B X_2 \to X_2 \wedge_B X_1$ is the switching equivalence.

Note that if E is a fibrewise Hopf space then the group

$$\pi_B^B(\Sigma_B^B(X_1 \times_B X_2), E)$$

is commutative, and so the fibrewise Whitehead product vanishes in this case. In any case the fibrewise Whitehead product vanishes under reduced fibrewise suspension, at least when X_1 and X_2 are fibrewise compact and fibrewise regular. For since the fibrewise loop-space $\Omega_B \Sigma_B^B(E)$ is fibrewise Hopf the fibrewise Whitehead product

$$\pi_B^B(\Sigma_B^B(X_1), \Omega_B \Sigma_B^B(E)) \times \pi_B^B(\Sigma_B^B(X_2), \Omega_B \Sigma_B^B(E))$$
$$\downarrow$$
$$\pi_B^B(\Sigma_B^B(X_1 \wedge_B X_2), \Omega_B \Sigma_B^B(E))$$

vanishes. If we now precompose with the product of the homomorphisms induced by the adjoint $E \to \Omega_B \Sigma_B^B(E)$ of the identity on E, and postcompose with the standard isomorphism

$$\pi_B^B(\Sigma_B^B(X \wedge_B X_2), \Omega_B \Sigma_B^B(E)) \to \pi_B^B(\Sigma_B^{B2}(X_1 \wedge_B X_2), \Sigma_B^B(E)),$$

we obtain the fibrewise suspension of the fibrewise Whitehead product

$$\pi_B^B(\Sigma_B^B(X_1), E) \times \pi_B^B(\Sigma_B^B(X_2), E) \to \pi_B^B(\Sigma_B^B(X_1 \wedge_B X_2), E),$$

which therefore vanishes as asserted.

Recall that in group theory, a given element k of a group G determines an automorphism of each normal subgroup H of G, by conjugation. We denote the automorphism thus: $h \to h^k$ $(h \in H)$, and use this notation in what follows, where for greater convenience we express the group operations in additive rather than multiplicative form although the groups in question are not, in general, commutative.

We now show† that the fibrewise Whitehead product is bilinear, provided the fibrewise cogroup-like spaces $\Sigma_B^B(X_i)$ $(i = 1, 2)$ are fibrewise

---

† For information about ordinary Whitehead products see [1] or [3].

† I am grateful to Dr John Rutter for advice as to the best way to deal with the arguments which follow.

homotopy-commutative as is the case when $X_i = \Sigma_B^B(Y_i)$ for some fibrewise pointed space Y_i.

Proposition (22.13). *Let X_i $(i = 1, 2)$ be a fibrewise non-degenerate space over B such that $\Sigma_B^B(X_i)$ is fibrewise homotopy-commutative. If $\alpha_i, \alpha_i' \in \pi_B^B(\Sigma_B^B(X_i), E)$, where E is a fibrewise pointed space, then*

$$[\alpha_1 + \alpha_1', \alpha_2] = [\alpha_1, \alpha_2] + [\alpha_1', \alpha_2],$$

$$[\alpha_1, \alpha_2 + \alpha_2'] = [\alpha_1, \alpha_2] + [\alpha_1, \alpha_2'].$$

In $\pi_B^B(\Sigma_B^B(X_1 \times_B X_2), E)$ the fibrewise Whitehead product, regarded as a commutator, satisfies the identity $[\alpha_1 + \alpha_1', \alpha_2] = [\alpha_2', \alpha_2]^{\alpha_1} + [\alpha_1, \alpha_2]$. Now the elements $[\alpha_1', \alpha_2]^{\alpha_1}$ and $[\alpha_1', \alpha_2]$ differ by the iterated commutator $[[\alpha_1', \alpha_2], -\alpha_1]^{\alpha_1}$. By exactness the latter lies in the normal subgroup $\pi_B^B(\Sigma_B^B(X_1 \wedge_B X_2), E)$, which is commutative since $\Sigma_B^B(X_1)$ is fibrewise homotopy-commutative and hence

$$(\Sigma_B^B(X_1)) \wedge_B X_2 = \Sigma_B^B(X_1 \wedge_B X_2)$$

is fibrewise homotopy-commutative. Thus the iterated commutator is zero and so we may drop the index α_1 in the above identity. This proves the first part of (22.13); the second part may be proved similarly or may be deduced from the first part by use of the commutation law (22.12).

A slightly more elaborate argument of the same type may be used to prove the Jacobi identity for the fibrewise Whitehead product as follows. Let X_i $(i = 1, 2, 3)$ be a fibrewise non-degenerate space such that $\Sigma_B^B(X_i)$ is fibrewise homotopy-commutative. Let $\alpha_i \in \pi_B^B(\Sigma_B^B(X_i), E)$ $(i = 1, 2, 3)$, where E is a fibrewise pointed space. I assert that

(22.14) $[[\alpha_1, \alpha_2], \alpha_3] + (\Sigma_B^B\tau)^*[[\alpha_2, \alpha_3], \alpha_1] + (\Sigma_B^B\tau^2)^*[[\alpha_3, \alpha_1], \alpha_2] = 0,$

where τ is the appropriate cyclic permutation of the factors of the fibrewise smash product.

To establish the identity we recall that the relation

$$[[-\alpha_2, \alpha_1], \alpha_3]^{\alpha_2} + [[-\alpha_3, \alpha_2], \alpha_1]^{\alpha_3} + [[-\alpha_1, \alpha_3], \alpha_2]^{\alpha_1} = 0$$

holds in

$$\pi_B^B(\Sigma_B^B(X_1 \times_B X_2 \times_B X_3), E)$$

(in fact holds in any group). Now the elements $[[-\alpha_2, \alpha_1], \alpha_3]^{\alpha_2}$ and $[[\alpha_1, \alpha_2], \alpha_3]]$ differ by the commutator

$$[[\alpha_2, \alpha_1], [-\alpha_3, \alpha_1]]^{\alpha_3}$$

Both $[\alpha_2, \alpha_1]$ and $[-\alpha_3, \alpha_1]$ here lie in the normal subgroup

$$\pi_B^B(\Sigma_B^B(X_1 \wedge_B (X_2 \times_B X_3)), E).$$

But $\Sigma_B^B(X_1 \wedge_B (X_2 \times_B X_3))$ is fibrewise homotopy-commutative, since $\Sigma_B^B(X_1)$ is fibrewise homotopy-commutative, and so the commutator vanishes since the subgroup is commutative. Thus the relation in

$$\pi_B^B(\Sigma_B^B(X_1 \times_B X_2 \times_B X_3), E)$$

reduces to

$$[[\alpha_1, \alpha_2], \alpha_3] + [[\alpha_2, \alpha_3], \alpha_1] + [[\alpha_3, \alpha_1], \alpha_2] = 0.$$

It only remains to translate this into terms of fibrewise Whitehead products in the group

$$\pi_B^B(\Sigma_B^B(X_1 \wedge_B X_2 \wedge_B X_2), E),$$

which has the effect of introducing the automorphisms in (22.14).

   This is as far as I propose to take the theory of fibrewise Whitehead products in the present volume, apart from some calculations which are given in §26. It is certainly possible to develop the theory further. For example, under suitable restrictions one can show that the fibrewise pointed space X over B is fibrewise Hopf if and only if the fibrewise Whitehead square of the class of the identity on $\Sigma_B^B(X)$ vanishes in $\pi_B^B(\Sigma_B^B(X \wedge_B X), \Sigma_B^B(X))$.

23. Fibrewise fibrations

We now turn from fibrewise cofibrations to fibrewise fibrations. As the terminology suggests the latter theory is to some extent dual to the former. However the duality is somewhat formal in character and, although often suggestive, does not extend to the deeper properties.

Definition (23.1). *The fibrewise map $\phi: E \to F$, where E and F are fibrewise spaces over B, is a fibrewise fibration if ϕ has the following property for all fibrewise spaces X. Let $f: X \to E$ be a fibrewise map and let $g: I \times X \to F$ be a fibrewise homotopy such that $g\sigma_0 = \phi f$. Then there exists a fibrewise homotopy $h: I \times X \to E$ such that $h\sigma_0 = f$ and $\phi h = g$.*

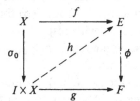

The property involved here is called the *fibrewise homotopy lifting property*; the fibrewise homotopy g of ϕf is lifted to a fibrewise homotopy

h of f itself. If $\phi: E \to F$ is a fibrewise fibration and F is fibrewise pointed with section t then the fibrewise pull-back of the triad

$$E \overset{\phi}{\to} F \overset{t}{\leftarrow} B$$

is called the *fibrewise fibre* of ϕ, with respect to t.

Clearly the composition of fibrewise fibrations is again a fibrewise fibration. Also the fibrewise product of fibrewise fibrations is a fibrewise fibration. It is easy to see that if $\phi: E \to F$ is a fibrewise fibration over B then the pull-back $\lambda^*\phi: \lambda^*E \to \lambda^*F$ is a fibrewise fibration over B' for each space B' and map $\lambda: B' \to B$.

Note that if ϕ is a fibration, in the ordinary sense, then ϕ is a fibrewise fibration. This is, of course, the opposite to the situation which occurs in the case of cofibrations. Also note that the projection $p: E \to B$ is always a fibrewise fibration for any fibrewise space E; thus there exist fibrewise maps which are fibrewise fibrations but not fibrations in the ordinary sense.

The proofs of our next two results are just formal exercises in the use of adjoints.

Proposition (23.2). *Let $\phi: E \to F$ be a fibrewise fibration, where E and F are fibrewise spaces over B. Then the postcomposition function*

$$\phi_*: \operatorname{map}_B(Y, E) \to \operatorname{map}_B(Y, F)$$

is a fibrewise fibration, for all fibrewise locally compact and fibrewise regular Y.

Proposition (23.3). *Let $u: A \to X$ be a fibrewise cofibration, where X is fibrewise locally compact and fibrewise regular over B and A is a closed subspace of X. Then the precomposition function*

$$u^*: \operatorname{map}_B(X, E) \to \operatorname{map}_B(A, E)$$

is a fibrewise fibration for all fibrewise spaces E.

Suppose, moreover, that E is fibrewise pointed and hence $\operatorname{map}_B(A, E)$ is fibrewise pointed. Then it follows from (9.4) that the fibrewise fibre of the fibrewise fibration in (23.3) with respect to the section of E can be identified with the fibrewise pointed space $\operatorname{map}_B^B(X/_B A, E)$ when the projection $X \to X/_B A$ is a proper map.

An important special case of (23.3) occurs when $X = I \times B$ and $A = * \times B$, with u mapping $(*, b)$ to $(0, b)$ for all $b \in B$. The result shows that $\rho_0: P_B(E) \to E$ is a fibrewise fibration for all fibrewise spaces E. More generally, let E' be a fibrewise space and let $\xi: E' \to E$ be a fibrewise map.

The *fibrewise mapping path-space* $W_B(\xi)$ of ξ is defined as the fibrewise pull-back of the triad

$$P_B E \xrightarrow{\rho_0} E \xleftarrow{\xi} E'.$$

Since $P_B(E)$ is a fibrewise fibration over E the fibrewise mapping path-space $\xi^* P_B(E) = W_B(\xi)$ is a fibrewise fibration over E'.

Changing the notation slightly, consider the fibrewise mapping path-space $W = W_B(\phi)$ of a fibrewise map $\phi\colon E \to F$, where E and F are fibrewise spaces over B. Clearly W is fibrewise contractible as a fibrewise space over E. Since W is the fibrewise pull-back of the triad

$$P_B(F) \xrightarrow{\rho_0} F \xleftarrow{\phi} E$$

there is a canonical fibrewise map k of $P_B(E)$ into W determined by the cotriad

$$P_B(F) \xleftarrow[P_B(\phi)]{} P_B(E) \xrightarrow[\rho_0]{} E$$

Proposition (23.4). *The fibrewise map $\phi\colon E \to F$ is a fibrewise fibration if and only if the fibrewise map $k\colon P_B(E) \to W$ admits a right inverse.*

For suppose that ϕ is a fibrewise fibration. Take W as the domain in the definition (23.1), take f to be π_2 and take g to be given by

$$I \times W \xrightarrow{\mathrm{id} \times \pi_1} I \times P_B(F) \xrightarrow{\alpha} F,$$

where α is fibrewise evaluation. We obtain a fibrewise homotopy $h\colon I \times W \to E$ of which the adjoint $\hat{h}\colon W \to P_B(E)$ is a right inverse of k.

Conversely suppose that $l\colon W \to P_B(E)$ is a right inverse of k. Let X be a fibrewise space, let $f\colon X \to E$ be a fibrewise map, and let $g_t\colon X \to F$ be a fibrewise homotopy such that $g_0 = \phi f$. Then a fibrewise map $\xi\colon X \to W$ is determined by the cotriad

$$P_B(F) \xleftarrow{\hat{g}} X \xrightarrow{f} E,$$

and the composition $l\xi\colon X \to P_B(E)$ is a lifting of $\hat{g}$ to a fibrewise homotopy of f, as required.

We already know that any fibrewise map $\phi\colon E \to F$ can be factored into the composition

$$E \xrightarrow{\sigma} W \xrightarrow{\rho_1} F,$$

where $W = W_B(\phi)$ and σ is a fibrewise homotopy equivalence. We are now going to show that ρ_1 here is a fibrewise fibration, so that ϕ can be factored into the composition of a fibrewise homotopy equivalence σ and a fibrewise fibration ρ_1.

In the proof we express fibrewise maps into W in the usual way by giving their components in E and $P_B(F)$; the latter we at once convert into fibrewise maps of the cylinder on the domain into F. So let X be a fibrewise space, let $f: X \to W$ be a fibrewise map, and let $g_t: X \to F$ be a fibrewise homotopy such that $g_0 = \rho_1 f = f'' \sigma_1$, where $f': X \to E$ and $f'': I \times X \to F$ are the components of f. Then a fibrewise map

$$h'': I \times I \times X \to F$$

is defined by

$$h''(s, t, x) = \begin{cases} f''(2s(2-t)^{-1}, x) & (s \leqslant 1 - t/2) \\ g(2s - 2 + t, x) & (s \geqslant 1 - t/2). \end{cases}$$

Take h'' as the second component of a fibrewise deformation h of f in which the first component h' remains stationary at f'. Then $h_0 = f$ and $\rho_1 h_t = g_t$, as required.

There are some important results due to Strøm in the ordinary case which combine features of fibrewise fibration theory and fibrewise cofibration theory. For example

Proposition (23.5). *Let $\phi: E \to F$ be a fibrewise fibration, where E and F are fibrewise spaces over B. Let F' be a closed subspace of F. If the pair (F, F') is fibrewise cofibred then so is the pair (E, E'), where $E' = \phi^{-1}F'$.*

For let $\alpha: F \to I$ and $h: I \times F \to F$ be a fibrewise Strøm structure for the pair (F, F'). Lift the fibrewise deformation $h_t \phi: E \to F$ of ϕ to a fibrewise deformation $k_t: E \to E$ of id_E. Then (β, l) is a fibrewise Strøm structure for the pair (E, E'), where

$$l(t, e) = k(\min(t, \alpha\phi(e)), e), \qquad \beta(e) = \min(2\alpha\phi(e), 1).$$

Proposition (23.6). *Let X be a fibrewise space over B and let $\alpha: X \to I$ be a map such that $A = \alpha^{-1}(0)$ is a fibrewise deformation retract of X. Let $\phi: E \to F$ be a fibrewise fibration, where E and F are fibrewise spaces. Then for each pair of fibrewise maps $f: A \to E$ and $g: X \to F$ such that $\phi f = g \,|\, A$ there exists a fibrewise extension $h: X \to E$ of f such that $\phi h = g$.*

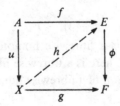

For let $r: X \to A$ be a fibrewise retraction and let H be a fibrewise homotopy rel A of ur into id_X. Define $D: I \times X \to X$ by

$$D(t, x) = H(\min(1, t/\alpha(x)), x).$$

Since ϕ is a fibrewise fibration there exists a fibrewise homotopy $K: I \times X \to E$ such that $\phi K = gD$ and such that $K(0, x) = fr(x)$. Then

$$h(x) = k(\alpha(x), x) \quad (x \in X)$$

defines a filler as required. Notice, incidentally, that h is unique up to fibrewise homotopy rel A. The main use of this result is to prove (23.7), which combines fibrewise homotopy extension and fibrewise homotopy lifting.

Proposition (23.7). *Let $\phi: E \to F$ be a fibrewise fibration, where E and F are fibrewise spaces over B. Let (X, A) be a closed fibrewise cofibred pair. Let $f: X \to E$ be a fibrewise map and let $g_t: X \to F$, $l_t: A \to E$ be fibrewise homotopies such that $\phi l_t = g_t | A$ and such that $g_0 = \phi f$, $l_0 = f | A$. Then there exists a fibrewise homotopy $h_t: X \to E$ of f extending l_t and lifting g_t.*

For $(\{0\} \times X) \cup (I \times A)$ is a fibrewise deformation retract of $I \times X$, as we have seen. Moreover there exists a map $\beta: X \to I$ such that $A = \beta^{-1}\{0\}$. Hence $\alpha^{-1}(0) = (\{0\} \times X) \cup (I \times A)$, where $\alpha: I \times X \to I$ is given by $\alpha(t, x) = t \cdot \beta(x)$. This proves (23.7).

We now turn to a series of results which the reader will at once recognize as duals of the corresponding results for fibrewise cofibrations given in §20.

Proposition (23.8). *Let E be a fibrewise space over B and let X and Y be fibrewise spaces over E. Let $\xi: X \to E$ and $\eta: Y \to E$ be fibrewise maps over E, and hence also over B. Let $\phi: X \to Y$ be a fibrewise map over B such that $\eta \phi \simeq_B \xi$. If η is a fibrewise fibration then $\phi \simeq_B \psi$ for some fibrewise map ψ over E such that $\eta \psi = \xi$.*

For let $g: I \times X \to E$ be a fibrewise homotopy, over B, of $\eta \phi$ into ξ. Since $g_0 = \eta \phi$ and since η is a fibrewise fibration there exists a lifting $h: I \times X \to Y$ of g such that $h_0 = \phi$. Take ψ to be h_1; then $\eta \psi = \xi$, as asserted.

Corollary (23.9). *Let $\xi: X \to E$ be a fibrewise fibration, where X and E are fibrewise spaces over B. If ξ admits a right inverse up to fibrewise homotopy then ξ admits a right inverse.*

Proposition (23.10). *Let $\xi\colon X\to E$ be a fibrewise fibration, where X and E are fibrewise spaces over B. Let $\theta\colon X\to X$ be a fibrewise map over E, and suppose that θ, as a fibrewise map over B, is fibrewise homotopic to the identity. Then there exists a fibrewise map $\theta'\colon X\to X$ over E such that $\theta\theta'$ is fibrewise homotopic to the identity over E.*

For let f_t be a fibrewise homotopy over B of θ into id_x. Then ξf_t is a fibrewise homotopy of ξ into itself, also over B. Since ξ is a fibrewise fibration there exists, by (23.8), a fibrewise homotopy $g_t\colon X\to E$ of id_X over ξf_t. Take $\theta'=g_1$; we shall prove that $\theta\theta'$ is fibrewise homotopic to the identity over E. For consider the juxtaposition k_s of θg_{1-s} and f_s, as a fibrewise homotopy over B of $\theta\theta'$ into the identity. Now $\xi f_t=\xi g_t$ and hence $H_{(s,0)}=\phi k_s$, where $H\colon I\times I\times X\to E$ is given by

$$H(s,t,x)=\begin{cases}\xi g(1-2s(1-t),x) & (0\leqslant s\leqslant\tfrac12)\\ \xi f(1-2(1-s)(1-t),x) & (\tfrac12\leqslant s\leqslant1).\end{cases}$$

Since ξ is a fibrewise fibration we can lift H to a fibrewise map $K\colon I\times I\times X\to X$ over E such that $K_{(s,0)}$ is fibrewise homotopic to k_s over E. Then

$$\theta\theta'=k_0\simeq K_{(0,0)}\simeq K_{(0,1)}\simeq K_{(1,1)}\simeq K_{(1,0)}\simeq k_1=\mathrm{id}_X,$$

all fibrewise homotopies being over E.

The main use of (23.10) is to prove

Proposition (23.11). *Let E be a fibrewise space over B and let X and Y be fibrewise spaces over E. Let $\xi\colon X\to E$ and $\eta\colon Y\to E$ be fibrewise fibrations. Let $\phi\colon X\to Y$ be a fibrewise map over E. Suppose that ϕ, as a fibrewise map over B, is a fibrewise homotopy equivalence. Then ϕ is a fibrewise homotopy equivalence over E.*

For let $\psi\colon Y\to X$ be a fibrewise homotopy inverse of ϕ, as a fibrewise map over B. Then $\xi\psi=\eta\phi\psi\simeq\eta$, by a fibrewise homotopy over B. Hence $\psi\simeq_B\psi'$ for some fibrewise map ψ' over E. Since $\phi\psi'\simeq_B\mathrm{id}_X$ and since $\phi\psi'$ is fibrewise over E there exists, by (23.10), a fibrewise map $\psi''\colon Y\to Y$ over E such that $\phi\psi'\psi''$ is fibrewise homotopic to the identity over E. Thus $\phi'=\psi'\psi''$ is a fibrewise homotopy right inverse of ϕ over E.

Now ϕ' is a fibrewise homotopy equivalence over B, since ϕ is a fibrewise homotopy equivalence over B, and so the same argument, applied to ϕ' instead of ϕ, shows that ϕ' admits a fibrewise homotopy right inverse ϕ'' over E. Thus ϕ' admits both a fibrewise homotopy left inverse ϕ over E and a fibrewise homotopy right inverse ϕ'' over E. Hence ϕ' is a fibrewise homotopy equivalence over E and so ϕ itself is a fibrewise homotopy equivalence over E, as asserted.

Corollary (23.12). *Let E be a fibrewise space over B. Let $\phi: X \to Y$ be a fibrewise fibration, where X and Y are fibrewise spaces over E. If ϕ is a fibrewise homotopy equivalence over B then the fibrewise mapping path-space $W_B(\phi)$ is fibrewise contractible over Y.*

Here we regard $W = W_B(\phi)$ as a fibrewise space over Y with projection the fibrewise fibration ρ_1. Now $\phi = \rho_1\sigma$, where $\sigma: X \to W$ is the standard embedding. Also ϕ and ρ_1 are fibrewise fibrations, and ρ_1 is a fibrewise homotopy equivalence over B, since ϕ and σ are fibrewise homotopy equivalences over B. Therefore ρ_1 is a fibrewise homotopy equivalence over Y, by (23.11), i.e. $W_B(\phi)$ is fibrewise contractible over Y, as asserted.

Corollary (23.13). *Let E be a fibrewise space over B and let $\xi: X \to E$ be a fibrewise fibration. Then the fibrewise path-space $P_B(X)$ is fibrewise contractible over the fibrewise mapping path-space $W_B(\xi)$.*

Of course the converse is obvious from (23.4). To prove (23.13) first observe that the projection $k: P_B(E) \to W_B(\xi)$ is a fibrewise fibration. This follows from the definition (23.1) and the observation that $(I \times I, I \times \{0\} \cup \{0\} \times I)$ is homeomorphic to $(I \times I, I \times \{0\})$. Then observe that k is a fibrewise homotopy equivalence over B and hence, by (23.11), a fibrewise homotopy equivalence over $W_B(\xi)$. This completes the proof.

We now come to a set of results which have no obvious parallel in fibrewise cofibration theory. Let A be a fibrewise space over B and let D be a fibrewise space over the cylinder $I \times A$. We regard D as a fibrewise space over A by composing with the projection $I \times A \to A$. We also regard $D_t = D \mid (\{t\} \times A)$ $(0 \leqslant t \leqslant 1)$ as a fibrewise space over A in the obvious way so that the inclusion $u_t: D_t \to D$ is a fibrewise map over A. We prove

Proposition (23.14). *Suppose that the projection $D \to I \times A$ is a fibrewise fibration over A. Then D_0 is a fibrewise deformation retract of D over A.*

For consider the fibrewise map $\xi: I \times I \times A \to I \times A$ which is given by $\xi(s, t, x) = ((1 - s)t, x)$. We can lift $\xi(\mathrm{id}_I \times \phi): I \times D \to I \times A$, where ϕ is the projection, to a fibrewise homotopy $h: I \times D \to D$, rel D_0, such that $h_0 = \mathrm{id}_D$. Now $h_1 D \subset D_0$, and so h_1 determines a fibrewise retraction $r_0: D \to D_0$ over A. We have $r_0 u_0 = \mathrm{id}_{D_0}$ and $u_0 r_0 = h_1 \simeq h_0 = \mathrm{id}_D$ by a fibrewise homotopy over A, as required.

Similarly D_1 is a fibrewise deformation retract of D over A, under the same hypothesis, and so D_0 has the same fibrewise homotopy type as D_1, over A. More specifically, there exists a fibrewise homotopy equivalence

$r_1 u_0 \colon D_0 \to D_1$ over A such that

$$u_1(r_1 u_0) = (u_1 r_1)u_0 \simeq u_0$$

by a fibrewise homotopy over A. We use this to obtain

Proposition (23.15). *Let E be a fibrewise space over B and let $\xi \colon X \to E$ be a fibrewise fibration. Let $\phi_0, \phi_1 \colon A \to E$ be fibrewise homotopic maps, where A is a fibrewise space over B. Then $\phi_0^* X$ and $\phi_1^* X$ have the same fibrewise homotopy type over A.*

In fact (23.15) follows at once from (23.14) on taking $D = \phi^* X$, where $\phi \colon I \times A \to X$ is a fibrewise homotopy of ϕ_0 into ϕ_1.

Corollary (23.16). *Let E be a fibrewise contractible fibrewise space over B and let X be a fibrewise fibration over E. Then X is trivial in the sense of fibrewise homotopy type over E.*

Proposition (23.17). *Let E be a fibrewise space over B and let Y, Z be fibrewise fibrations over E with projections η, ζ, respectively. If Y is fibrewise locally compact and fibrewise regular over E then $\operatorname{map}_E(Y, Z)$ is a fibrewise fibration over E.*

We work directly from the definition (23.1) of a fibrewise fibration. Let X be any fibrewise space and let $\phi \colon I \times X \to E$ be a fibrewise homotopy. Let

$$\hat{f} \colon X \to \operatorname{map}_E(Y, Z)$$

be a fibrewise map over ϕ_0. Since Y is fibrewise locally compact and fibrewise regular over E the adjoint

$$f \colon X \times_E Y \to Z$$

of $\hat{f}$ is defined, where X is regarded as a fibrewise space over E through ϕ_0. Now consider the fibrewise topological product

$$(I \times X) \times_E Y,$$

where $I \times X$ is regarded as a fibrewise space over E through ϕ. Recall that $X \times_E Y$ is a fibrewise deformation retract of $(I \times X) \times_E Y$ over X, where X is embedded in $I \times X$ through u_0. If

$$r_0 \colon (I \times X) \times_E Y \to X \times_E Y$$

is a fibrewise retraction then

$$f r_0 \colon (I \times X) \times_E Y \to Z$$

agrees with f on $X \times_E Y$. Now let $h_t \colon I \times X \to I \times X$ be defined by $h_t(s, x) = (st, x)$. Then

$$(I \times X) \times_E Y \overset{\pi_1}{\to} I \times X \overset{h_t}{\to} I \times X \overset{\phi}{\to} E$$

is a fibrewise homotopy of $\zeta f r_0$ into $\phi \pi_1$, rel$(X \times_E Y)$. Since $I \times X, X)$ is always a fibrewise cofibration we can lift this to a fibrewise homotopy, rel $(X \times_E Y)$, of $f r_0$ into a fibrewise map $g \colon (I \times X) \times_E Y \to Z$ over $\phi \pi_1$. Now the adjoint

$$\hat{g} \colon I \times X \to \mathrm{map}_E(Y, Z)$$

is over ϕ and satisfies $\hat{g} u_0 = \hat{f}$. Therefore $\mathrm{map}_E(Y, Z)$ is a fibrewise fibration over E, as asserted.

To illustrate some of the ideas of this section let us prove a fibrewise version of the well-known Spanier–Whitehead theorem of [57], as follows.

Proposition (23.18). *Let* $\phi \colon E \to F$ *be a fibrewise pointed map, where E and F are fibrewise pointed spaces over B. Suppose that ϕ is a fibrewise fibration and that the fibrewise fibre D is fibrewise well-pointed. Also suppose that D is fibrewise pointed contractible in E. Then D is a fibrewise Hopf space.*

Here we say that D is fibrewise pointed contractible in E if the inclusion $u \colon D \to E$ is fibrewise pointed nulhomotopic. Choose a fibrewise pointed homotopy $u_t \colon D \to E$ of the fibrewise constant into u. Since the fibrewise pair $(D \times_B D, D \vee_B D)$ is fibrewise cofibred by (20.17) we may extend

$$u_t \vee u \colon D \vee_B D \to E$$

to a fibrewise pointed homotopy

$$g_t \colon D \times_B D \to E,$$

rel$(B \times_B D)$, of $u \pi_2$ into a fibrewise pointed map $k \colon D \times_B D \to E$, where $k(D \vee_B D) \subset D$. I assert that there exists a fibrewise pointed homotopy h_t of k such that $h_t(D \vee_B D) \subset D$ for all values of the parameter t and such that $h_1(D \times_B D) \subset D$.

For consider the fibrewise pointed homotopy $g'_t \colon D \times_B D \to E$ given by

$$g'_t = \begin{cases} g_{1-3t} & (0 \leqslant t \leqslant 1/3) \\ u_{3t-1} \pi_2 & (1/3 \leqslant t \leqslant 2/3) \\ u_{3t-2} \pi_1 & (2/3 \leqslant t \leqslant 1). \end{cases}$$

Clearly $g' \,|\, I \times (D \vee_B D)$ is fibrewise pointed homotopic to the composition

$$I \times (D \vee_B D) \underset{\pi_2}{\to} D \vee_B D \underset{\nabla}{\to} D \underset{u}{\to} E,$$

rel$(\dot{I} \times (D \times_B B) \cup I \times (B \times_B D))$. Since the pair

$$(I \times (D \times_B D), \dot{I} \times (D \times_B B) \cup I \times (B \times_B D))$$

is fibrewise cofibred, by (20.17), the existence of h_t follows at once from the fibrewise homotopy extension property.

Now $\phi \circ h_t$ is a fibrewise pointed nulhomotopy, rel($D \vee_B D$), of $\phi \circ k \colon D \times_B D \to F$. Since $(D \times_B D, D \vee_B D)$ is fibrewise cofibred, and since ϕ is a fibrewise fibration, it follows from (23.7) that the fibrewise constant map $D \times_B D \to F$ can be lifted to a fibrewise pointed map $D \times_B D \to E$ which agrees with $u \vee u$ on $D \vee_B D$. Thus D is a fibrewise Hopf space, as asserted.

24. Relation with equivariant homotopy theory

Let G be a topological group. Recall that a *G-homotopy* between G-maps is just a homotopy in the ordinary sense which is equivariant at every stage of the deformation. Specifically, a G-homotopy between the G-maps $\theta, \phi \colon X \to Y$, where X and Y are G-spaces, is a homotopy $f_t \colon X \to Y$ of θ into ϕ such that f_t is a G-map for all t. Note that a G-homotopy f_t determines a homotopy $(f/G)_t \colon X/G \to Y/G$. A G-homotopy into a constant G-map is called a *G-nulhomotopy*; of course the constant is necessarily a fixed point.

The proof that the existence of a G-homotopy constitutes an equivalence relation $\simeq_G$ between G-maps is essentially the same as in the non-equivariant theory. The set of equivalence classes of G-maps of the G-space X into the G-space Y is denoted by $\pi_G(X, Y)$. The operation of composition for G-maps induces an operation of composition for G-homotopy classes

$$\pi_G(Y, Z) \times \pi_G(X, Y) \to \pi_G(X, Z).$$

Thus we may, when convenient, pass from the category of G-spaces and G-maps to the category of G-spaces and G-homotopy classes of G-maps.

The G-maps which induce equivalences in the latter category are called *G-homotopy equivalences*, and G-spaces which are equivalent in this sense are said to have the same *G-homotopy type*. A G-space of the same G-homotopy type as the point-space is said to be *G-contractible*. For example the cone $\Gamma(X)$ is G-contractible for every G-space X. Clearly X is G-contractible if and only if the identity id_X is G-nulhomotopic. A G-space which is contractible, as an ordinary space, is not necessarily G-contractible. For example take the real line $\mathbb{R}$, regarded as a $\mathbb{Z}$-space in the standard way so that $\mathbb{R}/\mathbb{Z}$ is the circle group; although $\mathbb{R}$ is contractible there is not even a fixed point. Note that the G-homotopy type of the G-space X determines the homotopy type of the orbit space X/G.

Let $\theta: X \to Y$ and $\phi: Y \to X$ be G-maps such that $\phi\theta$ is G-homotopic to id_X, where X and Y are G-spaces. Then ϕ is a *left inverse* of θ and θ is a *right inverse* of ϕ, up to G-homotopy. Note that if θ admits a left inverse ϕ and a right inverse ϕ', up to G-homotopy, then ϕ and ϕ' are G-homotopic and so θ is a G-homotopy equivalence.

From a formal point of view π_G constitutes a binary functor from the category of G-spaces to the category of sets, contravariant in the first entry and covariant in the second. There is a natural equivalence

$$\pi_G(X, Y_1 \times Y_2) \to \pi_G(X, Y_1) \times \pi_G(X, Y_2),$$

for all G-spaces X, Y_1, Y_2, and a natural equivalence

$$\pi_G(X_1 + X_2, Y) \to \pi_G(X_1, Y) \times \pi_G(X_2, Y),$$

for all G-spaces X_1, X_2, Y. Clearly the G-homotopy types of $X \times Y$ and $X + Y$ depend only on the G-homotopy types of X and Y.

If E is a G-space then postcomposition with $g_\#: E \to E$ $(g \in G)$ determines an action of G on the path-space $P(E)$, which may therefore be regarded as a G-space. We use this in the definition of G-cofibration, as follows.

Definition (24.1). *The G-map $u: A \to X$, where A and X are G-spaces, is a G-cofibration if u has the following G-homotopy extension property. Let $f: X \to E$ be a G-map, where E is a G-space, and let $g_t: A \to P(E)$ be a G-homotopy of $f \mid A$. Then g_t can be extended to a G-homotopy $h_t: X \to E$ of f.*

The theory of G-cofibrations runs closely parallel to that of ordinary cofibrations. In particular, by using the same procedure as in §20, we obtain

Proposition (24.2). *Let (X, A) be a closed G-pair. Then (X, A) is G-cofibred if and only if $(\{0\} \times X) \cup (I \times A)$ is a G-retract of $I \times X$.*

By saying that (X, A) is G-cofibred here we mean, of course, that the inclusion $A \to X$ is a G-cofibration. Note that if (X, A) is a closed cofibred G-pair, with G compact, then $(X/G, A/G)$ is a closed cofibred pair.

We can also study G-cofibrations via Strøm G-structures. Specifically, let (X, A) be a closed G-pair. Then a *Strøm G-structure* on (X, A) is a Strøm structure (α, h) in the ordinary sense for which α is invariant and h is equivariant. As in the non-equivariant theory we find that (X, A) is a G-cofibration if and only if (X, A) admits a Strøm G-structure. Note that a Strøm G-structure (α, h) on (X, A), with G compact, determines a Strøm structure $(\alpha/G, h/G)$ on $(X/G, A/G)$.

For an example of a G-cofibration consider the closed pair (B^n, S^{n-1}) ($n \geqslant 1$) with $O(n)$ acting in the usual way. The standard retraction

$$I \times B^n \to (\{0\} \times B^n) \cup (I \times S^{n-1})$$

is an $O(n)$-map, and so (B^n, S^{n-1}) is $O(n)$-cofibred.

Among the properties of G-cofibred pairs the following is relevant.

Proposition (24.3). *Let* (X, A) *be a cofibred G-pair. Suppose that A is G-contractible. Then the natural projection from X to X/A is a G-homotopy equivalence.*

Recall from §7 that if T is a G-space and T_0 is a closed invariant subspace then an endofunctor Φ of the category of G-spaces is defined which transforms each G-space X into the equivariant push-out $\Phi(X)$ of the cotriad

$$T \times X \leftarrow T_0 \times X \to T_0$$

and similarly for G-maps. In particular suppose that X is a pointed G-space and that the basepoint x_0 is such that (X, x_0) is a cofibred G-pair. Then the pair $(\Phi(X), T)$ is also G-cofibred. Hence if T is G-contractible it follows from (24.3) that the natural projection $\Phi(X) \to \Phi^*(X) = \Phi(X)/T$ is a G-homotopy equivalence. In particular the natural projections $\Gamma(X) \to \Gamma^*(X)$ and $\Sigma(X) \to \Sigma^*(X)$ are G-homotopy equivalences; as a consequence we see that $\Gamma^*(X)$ is G-contractible.

We now turn to the pointed equivariant theory. Recall that a *pointed G-homotopy* between pointed G-maps is just a G-homotopy which respects basepoints. Specifically a pointed G-homotopy between pointed G-maps $\theta, \phi: X \to Y$, where X and Y are pointed G-spaces, is a homotopy $f_t: X \to Y$ of θ into ϕ which is both pointed and equivariant for all t. Note that a pointed G-homotopy f_t determines a pointed homotopy $f_t/G: X/G \to Y/G$. A pointed G-homotopy into the constant G-map is called a *pointed G-nulhomotopy*.

The proof that the existence of a pointed G-homotopy constitutes an equivalence relation $\simeq_G^*$ between pointed G-maps is just the same as in the non-equivariant theory. The set of equivalence classes of pointed G-maps of the pointed G-space X into the pointed G-space Y is denoted by $\pi_G^*(X, Y)$. The operation of composition for pointed G-maps induces an operation of composition for pointed G-homotopy classes:

$$\pi_G^*(Y, Z) \times \pi_G^*(X, Y) \to \pi_G^*(X, Z).$$

Thus we may, when convenient, pass from the category of pointed G-spaces and pointed G-maps to the category of pointed G-spaces and pointed G-homotopy classes of pointed G-maps.

The pointed G-maps which induce equivalences in the latter category are called *pointed G-homotopy equivalences*, and pointed G-spaces which are equivalent in this sense are said to have the same *pointed G-homotopy type*. A pointed G-space of the same pointed G-homotopy type as the point-space is said to be *pointed G-contractible*. Clearly a pointed G-space X is pointed G-contractible if and only if the identity id_X is pointed G-nulhomotopic. Note that the pointed G-homotopy type of the pointed G-space X determines the pointed homotopy type of the orbit space X/G.

From the formal point of view π_G^* constitutes a binary functor from the category of pointed G-spaces to the category of pointed sets, contravariant in the first entry and covariant in the second. There is a natural equivalence

$$\pi_G^*(X, Y_1 \times Y_2) \to \pi_G^*(X, Y_1) \times \pi_G^*(X, Y_2),$$

for all pointed G-spaces X, Y_1, Y_2 and a natural equivalence

$$\pi_G^*(X_1 \vee X_2, Y) \to \pi_G^*(X_1, Y) \times \pi_G^*(X_2, Y)$$

for all pointed G-spaces X_1, X_2, Y. Clearly the pointed G-homotopy types of $X \times Y$ and $X \vee Y$ depend only on the pointed G-homotopy types of X and Y.

A pointed G-space X is said to be a *group G-space* if X is equipped with pointed G-maps

$$m: X \times X \to X, \qquad v: X \to X$$

(multiplication and inversion) such that the group axioms are satisfied. For example, G itself may be regarded as a group G-space with the conjugation action. If the group axioms are satisfied up to pointed G-homotopy we describe X as a *group-like G-space*. The definition of cogroup-like G-space is analogous. For example if Y is a pointed G-space then the reduced suspension $\Sigma^*(Y)$ is a cogroup-like G-space, with comultiplication given by the track sum. Note that if X is a group-like G-space then the orbit space X/G is a group-like space, while if X is a cogroup-like G-space then X/G is a cogroup-like space.

The definition of pointed G-cofibration is analogous to that of G-cofibration, with G-spaces replaced by pointed G-spaces and so forth. Again a necessary and sufficient condition can be obtained involving pointed G-retraction. This implies that if (X, A) is a closed pointed G-cofibred pair then $(X/G, A/G)$ is a pointed cofibred pair.

Among the properties of pointed G-cofibred pairs the following is relevant.

Proposition (24.4). *Let (X, A) be a pointed G-cofibred pair. Suppose that A is pointed G-contractible. Then the natural projection from X to X/A is a pointed G-homotopy equivalence.*

The development of the equivariant version of the Puppe theory presents no particular problems. Thus starting from a pointed G-map $\phi: X \to Y$, where X and Y are pointed G-spaces, we derive a sequence

$$X \xrightarrow{\phi} Y \xrightarrow{\phi'} \Gamma^*(\phi) \xrightarrow{\phi''} \Sigma^*(X) \to \cdots$$

of pointed G-maps which is exact in the sense that the sequence

$$\pi_G^*(X, E) \xleftarrow{\phi^*} \pi_G^*(Y, E) \xleftarrow{\phi'^*} \pi_G^*(\Gamma^*(\phi), E) \xleftarrow{\phi''^*} \pi_G^*(\Sigma^*(X), E) \leftarrow \cdots$$

of pointed sets and pointed functions is exact for all pointed G-spaces E.

If X is a pointed G-space with basepoint x_0 the mapping cylinder $\check{X}$ of the inclusion $x_0 \to X$ is regarded as a pointed G-space, in the obvious way. The natural projection $\rho: \check{X} \to X$ is both a pointed G-map and a G-homotopy equivalence.

Definition (24.5). *The pointed G-space X is non-degenerate if ρ is a pointed G-homotopy equivalence.*

Note that if X is non-degenerate in the equivariant sense then X/G is non-degenerate in the non-equivariant sense. Clearly $\check{X}$ itself is always non-degenerate; thus every pointed G-space has the same G-homotopy type as a non-degenerate G-space. Also X is non-degenerate if (X, x_0) is a cofibred G-pair.

By a *Puppe G-structure* on the pointed G-space X I mean a Puppe structure (α, U), as in the non-equivariant theory, where α is invariant and U is G-contractible in X. The argument used in §22 shows that X is non-degenerate if and only if X admits a Puppe G-structure. Note that a Puppe G-structure (α, U) on X determines a Puppe structure $(\alpha/G, U/G)$ on X/G.

Proceeding just as in the ordinary theory we see that if X and Y are non-degenerate G-spaces then the pair $(X \times Y, X \vee Y)$ is G-cofibred and the smash product $X \wedge Y$ is non-degenerate in the equivariant sense. Moreover the smash product of non-degenerate G-spaces is G-homotopy-associative, by the argument used in §22. This leads to the conclusion that if X and Y are non-degenerate G-spaces then $\Sigma^*(X \times Y)$ has the same pointed G-homotopy type as the pointed coproduct

$$\Sigma^*(X) \vee \Sigma^*(Y) \vee \Sigma^*(X \wedge Y)$$

and similarly for products of more than two factors.

The equivariant version of the Whitehead product is defined by the same procedure as in §22. Thus let X_i $(i = 1, 2)$ be non-degenerate G-spaces. Then the equivariant Whitehead product constitutes a pairing

$$\pi_G^*(\Sigma^*(X_1), E) \times \pi_G^*(\Sigma^*(X_2), E) \rightarrow \pi_G^*(\Sigma^*(X_1 \wedge X_2), E)$$

for all pointed G-spaces E. The pairing is denoted by square brackets $[\ ,\ ]$ and satisfies the condition

$$[\alpha_1, \alpha_2] + (\Sigma^*\tau)^*[\alpha_2, \alpha_1] = 0,$$

where $\alpha_i \in \pi_G^*(\Sigma^*(X_i), E)$ $(i = 1, 2)$ and where τ denotes the switching equivalence between the smash products.

If the cogroup-like structure on $\Sigma^*(X_i)$ $(i = 1, 2)$ is G-homotopy-commutative, as is the case when X_i is of the form $\Sigma^*(Y_i)$ for $i = 1, 2$, then the pairing is bilinear, by the same argument as in §22. Similarly if the cogroup-like structure on $\Sigma^*(X_i)$ $(i = 1, 2, 3)$ is G-homotopy-commutative then the Jacobi identity is satisfied by the equivariant Whitehead product.

We are now ready to describe further the connection between the equivariant theory and the fibrewise theory. Consider, as before, a G-space P with orbit space $B = P/G$. The mixed product functor $P_\#$ from the category of G-spaces to the category of fibrewise spaces over B respects homotopies and thereby induces a function

$$P_\# : \pi_G(X, Y) \rightarrow \pi_B(P_\# X, P_\# Y),$$

where X and Y are G-spaces. Moreover, if X and Y have the same G-homotopy type then $P_\# X$ and $P_\# Y$ have the same fibrewise homotopy type, in particular if X is G-contractible then $P_\# X$ is fibrewise contractible. Finally by inspecting the retraction conditions we have

Proposition (24.6). *If* $u: A \rightarrow X$ *is a G-cofibration, where A and X are G-spaces, then $P_\# A \rightarrow P_\# X$ is a fibrewise cofibration over B.*

Turning now to the pointed theories, we observe that $P_\#$, regarded as a functor from the category of pointed G-spaces to the category of fibrewise pointed spaces over B, also respects homotopies, and thereby induces a function

$$P_\# : \pi_G^*(X, Y) \rightarrow \pi_B^B(P_\# X, P_\# Y),$$

where X and Y are pointed G-spaces. Moreover if X and Y have the same pointed G-homotopy type then $P_\# X$ and $P_\# Y$ have the same fibrewise pointed homotopy type, in particular if X is pointed G-contractible than $P_\# X$ is fibrewise pointed contractible.

Note that if X is a group-like G-space, with multiplication m, then $P_\# X$ is a fibrewise group-like space, with multiplication

$$P_\# m: P_\# X \times_B P_\# X \to P_\# X,$$

and that the binary system $(P_\# X, P_\# m)$ is homotopy-commutative, homotopy-associative etc. in the fibrewise sense if the binary system (X, m) is homotopy-commutative, homotopy-associative etc. in the equivariant sense. The dual situation is similar, with equivariant cogroup-like spaces transforming into fibrewise cogroup-like spaces, and so on.

By using the appropriate retraction conditions we obtain

Proposition (24.7). *Let $u: A \to X$ be a pointed G-cofibration, where A and X are pointed G-spaces. Then $P_\# u: P_\# A \to P_\# X$ is a fibrewise pointed cofibration over B.*

By using Puppe structure we also obtain

Proposition (24.8). *Let X be a non-degenerate G-space. Then $P_\# X$ is a fibrewise non-degenerate space over B.*

In fact if (α, U) is a Puppe G-structure on X then $(P_\# \alpha, P_\# U)$ is a fibrewise Puppe structure on $P_\# X$.

In conclusion let us compare the Whitehead products in the two theories. Let X_i $(i = 1, 2)$ be a non-degenerate G-space so that $E_i = P_\# X_i$ is a fibrewise non-degenerate space over B. Then for all pointed G-spaces X we have a commutative diagram as shown below, in which $E = P_\#(X)$ where the upper horizontal is the equivariant Whitehead product and where the lower horizontal is the fibrewise Whitehead product.

$$\begin{CD}
\pi_G^*(\Sigma^*(X_1), X) \times \pi_G^*(\Sigma^*(X_2), X) @>>> \pi_G^*(\Sigma^*(X_1 \wedge X_2), X) \\
@V{P_\# \times P_\#}VV @VV{P_\#}V \\
\pi_B^B(\Sigma_B^B(E_1), E) \times \pi_B^B(\Sigma_B^B(E_2), E) @>>> \pi_B^B(\Sigma_B^B(E_1 \wedge_B E_2), E)
\end{CD}$$

The relationship exhibited here will be used in §26 to obtain further information about fibrewise Whitehead products in certain cases.

Exercises

1. The subspace X of $B \times \mathbb{R}^n$ is *fibrewise star-like* in the sense that there exists a section $s: B \to X$ such that the line segment

$$(b, (1-t)s(b) + tx) \quad (0 \leqslant t \leqslant 1)$$

is entirely contained within X_b for each point $x \in X_b$, $b \in B$. Show that X is fibrewise contractible.

2. Let X be a fibrewise neighbourhood retract of $B \times \mathbb{R}^n$ for some n. Show that there exists a neighbourhood N of the diagonal ΔX in $X \times_B X$ such that the projections $\pi_i \colon X \times_B X \to X$ $(i = 1, 2)$ are fibrewise homotopic rel ΔX over N.

3. Let $u \colon A \to X$ be a fibrewise cofibration, where A and X are fibrewise spaces over B. Show that:

(i) if u admits a left inverse up to fibrewise homotopy then u admits a left inverse, as a fibrewise map;

(ii) if u admits a left inverse, as a fibrewise map, and is a fibrewise homotopy equivalence then u is a fibrewise homotopy equivalence under A.

4. Let (X, A) be a closed fibrewise cofibred pair. Suppose that $\Sigma_B^B(A)$ is a fibrewise retract of $\Sigma_B^B(X)$. Show that $\Sigma_B^B(X)$ has the same fibrewise homotopy type as the fibrewise pointed coproduct $\Sigma_B^B(A) \vee_B \Sigma_B^B(X/_B A)$.

5. The fibrewise map $u \colon A \to X$ is said to be a *weak fibrewise cofibration* if there exists a fibrewise cofibration $v \colon A \to Y$ such that X and Y have the same fibrewise homotopy type as fibrewise spaces under A. Show that if u has this property then for each fibrewise space E, fibrewise map $f \colon X \to E$ and fibrewise homotopy $g_t \colon A \to X$ of fu there exists a fibrewise homotopy $h_t \colon X \to E$ such that $h_t u = g_t$ and such that h_0 is fibrewise homotopic to f.

6. Show that (20.12) remains true when u is just a weak fibrewise cofibration.

V
Miscellaneous topics

---

25. Fibre bundles

This is not the place for a lengthy discussion of the theory of fibre bundles; after all, a full-scale treatment can be found in several of the well-known textbooks. It is, however, necessary to give a brief sketch so as to be in a position to show how the material presented earlier in this volume fits into the theory.

There are two main ways to approach the theory of fibre bundles. The first may be called the Whitney approach, since it goes back to the original paper of Whitney on 'sphere-spaces'. This fits in well with the requirements of differential topology. The other approach is due to H. Cartan and fits in well with the theory of topological transformation groups. Students of the subject need to understand both approaches and the relation between them.

As before we work over a topological base space B. In the crudest form of the definition a fibre bundle over B is a fibrewise topological space X which is locally trivial, in the sense of §1. Specifically there exists an open covering $\{B_j\}$ of B such that the restriction $X_j = X_{B_j}$ of X to B_j is trivial over B_j for each index j. Thus for each j we have a fibrewise topological equivalence $\phi_j: X_j \to B_j \times T_j$ for some topological T_j. To simplify matters we assume that T_j is independent of j, as can always be arranged when B is connected. Thus $\phi_j: X_j \to B_j \times T$ for some topological T, called the fibre. Note that $\phi_j\phi_k^{-1}$ determines a fibrewise topological equivalence of $(B_j \cap B_k) \times T$ with itself for each pair of indices j, k.

Even with this crude form of the definition several useful observations can be made. For example X is fibrewise open. If T is compact then X is fibrewise compact.

After these preliminary remarks we are ready to give the full definition of the concept of fibre bundle. We continue to work over the topological space B but now we also take a topological group G and a G-space T. We suppose an open covering $\{B_j\}$ of B and a family $\{g_{jk}\}$ of continuous functions $g_{jk}: B_j \cap B_k \to G$ to be given; here j and k run through the indexing set J of the covering. Finally we suppose that these functions satisfy the cocycle condition

$$g_{ik} = g_{ij} \cdot g_{jk} \quad (i, j, k \in J);$$

of course, the relation only holds on $B_i \cap B_j \cap B_k$. With such data a fibre bundle over B can be constructed as follows.

Consider the topological product $B \times T \times J$, where J has the discrete topology. Let X' denote the subspace of triples (b, t, j) such that $b \in B_j$ and $t \in T$. We regard X' as a fibrewise topological space over B using the first projection. By virtue of the cocycle condition a fibrewise equivalence relation R on X' is given by: $(b, t, j)R(b, u, k)$ if and only if $b \in B_j \cap B_k$ and $u = g_{jk}(b) \cdot t$. Then the fibrewise quotient space $X = X'/R$ is locally trivial; in fact a trivialization $\phi_j: X_j \to B_j \times T$ over B_j is defined so that $\phi_j^{-1}(b, t)$ is the equivalence class of (b, t, j). Thus X is a fibre bundle over B in the crude sense.

The procedure we have just described is functorial in the following sense. Let $\phi: T_1 \to T_2$ be a G-map, where T_1 and T_2 are G-spaces. Then

$$\text{id} \times \phi \times \text{id}: B \times T_1 \times J \to B \times T_2 \times J$$

transforms X_1' into X_2' and respects the equivalence relations R_1 and R_2 determined by the structural data. Since the projections are open the restriction of $\text{id} \times \phi \times \text{id}$ induces a continuous fibrewise function $\psi: X_1 \to X_2$. We refer to ψ as the function determined by ϕ.

An important special case is when G itself is the fibre. We regard G as a G-space using right translation, and then the procedure we have described produces a fibre bundle P over B with fibre G, called the principal G-bundle. Note that an action of G on the right of $B \times G \times J$ is given by

$$(b, t, j) \cdot g = (b, t \cdot g, j),$$

where $g, t \in G$, $b \in B$ and $j \in J$. The action leaves the subspace invariant; moreover the relation R is equivariant with respect to the action. Since the natural projection is open we obtain an induced action on P itself, called the principal action. Clearly the principal action is free, hence the function

$$\theta: P \times G \to P \times_B P$$

is bijective, where $\theta(x, g) = (x, x \cdot g)$. Obviously θ is continuous: in fact θ^{-1} is also continuous, by local triviality, and so θ is a homeomorphism.

In the Cartan approach† the procedures are essentially reversed. We begin with a topological group G and a free (right) G-space P. We regard

---

† For more details of the Cartan approach see [50].

P as a fibrewise topological space over the orbit space $B = P/G$, in the natural way. We suppose that the function

$$\theta: P \times G \to P \times_B P$$

is a homeomorphism, as is always the case when G is compact. We also suppose that P is locally sectionable over B. Then it is not difficult to show using (7.9) that P is a principal G-bundle over B in the previous sense. Moreover if T is a G-space then the mixed product $P \times_G T$ is a fibre bundle over B with fibre T.

An important special case is when G is discrete. In this case a free proper action is said to be *properly discontinuous*. If the action of G on P satisfies this condition then the mixed product $P \times_G T$ is a fibre bundle over B for each discrete G-space T. Such fibre bundles are called covering spaces. Incidentally the overlayings of Fox considered in §1 are always covering spaces, the converse holds for finite G.

When we turn from topological spaces to uniform spaces it is not immediately obvious whether either of the above approaches can be adapted to the new situation. The main purpose of this section is to establish that the Whitney approach can be so adapted; I do not know whether the same is true of the Cartan approach.

As before we work over a topological base space B. In the crudest form of the definition a uniform fibre bundle over B is a fibrewise uniform space X over B which is locally trivial in the sense of §12. Specifically there exists an open covering $\{B_j\}$ of B such that the restriction X_j of X to B_j is trivial over B_j for each index j. Thus for each j we have a fibrewise uniform equivalence $\phi_j: X_j \to B_j \times T_j$ for some uniform T_j. To simplify matters we assume that T_j is independent of j, as can always be arranged when B is connected. Thus $\phi_j: X_j \to B_j \times T$ for some uniform T, called the fibre. Note that $\phi_j \phi_k^{-1}$ determines a fibrewise uniform equivalence of $(B_j \cap B_k) \times T$ with itself for each pair of indices j, k.

Even with this crude form of the definition several useful observations can be made. For example if T is complete then X is fibrewise complete. Results such as these are special cases of results proved in Chapter III.

Before giving the full definition of the concept of uniform bundle we need to say what we mean for a topological group G to act uniformly on a uniform space X. It is not sufficient, for our purposes, to require the action to be continuous, with respect to the uniform topology, and for the translations to be uniform equivalences. On the other hand to require that the action $G \times X \to X$ should be uniformly continuous, when G itself has uniform structure, would be much too restrictive. The appropriate concept turns out to be as in

Definition (25.1). *Suppose that the topological group G acts on the uniform space X through uniform equivalences. The action is* uniform *if for each entourage D of X there exists a neighbourhood W of e such that* $(g\xi, \xi) \in D$ *for all* $g \in W$ *and* $\xi \in X$.

Note that if G acts uniformly on X then G acts uniformly on each invariant subspace of X. When G is discrete any action of G on X through uniform equivalences is necessarily uniform.

Of course it is sufficient for the condition in (25.1) to be satisfied for all members D of a uniform basis. For example, take $X = G$, acting on itself by right translation. The right uniformity is generated by the subsets

$$D_V = \{(\xi, \eta) \in G \times G: \xi\eta^{-1} \in V\},$$

where V runs through the neighbourhoods of e. In this case it is sufficient to take $W = V$ in order to satisfy the condition.

Proposition (25.2). *Suppose that the topological group G acts on the uniform space X through uniform equivalences. The action is uniform if and only if the function* $\phi: G \times X \to G \times X$ *is fibrewise uniformly continuous, where*

$$\phi(g, x) = (g, gx) \quad (g \in G, x \in X).$$

Here $G \times X$ is regarded as a fibrewise uniform space over G in the standard way, using the first projection. Obviously the action is uniform if the function is fibrewise uniformly continuous. Conversely, suppose that the action is uniform. Let D be an entourage of X and let g be an element of G. Then $g_{\#}^{-2}D$ is an entourage of X, since $g_{\#}$ is a uniform equivalence and so there exists a symmetric entourage E of X such that $E \circ E \circ E \subset g_{\#}^{-2}D$. Since the action is uniform there exists a neighbourhood W of e such that $(h\xi, \xi) \in E$ for all $h \in W$ and $\xi \in X$. Thus if $h, k \in W$ and $(\xi, \eta) \in E$ then $(h\xi, k\eta) \in E \circ E \circ E$ and so $(gh\xi, gk\eta) \in D$. Since gW is a neighbourhood of g this shows that ϕ is fibrewise uniformly continuous, as asserted.

Hence it follows using (17.2) that a topological action on a compact space X is necessarily uniform, with respect to the unique uniform structure. Thus the standard actions of the orthogonal group $O(n)$ ($n = 1, 2, \ldots$) on the closed n-ball B^n and on the $(n-1)$-sphere S^{n-1} are uniform. The action of $O(n)$ on the open n-ball is also uniform since Int B^n is an invariant subspace. Of course these results can also be proved directly. On the other hand the action of $O(n)$ on $\mathbb{R}^n$ itself is not uniform, as can be shown without difficulty.

For compact Hausdorff G a topological action on a locally compact Hausdorff X can be extended to a topological action on the Alexandroff compactification X^+. Since the latter is uniform, as we have just observed, so is the former, provided that the uniform structure of X is induced by that of X^+.

After these preliminary remarks we are now ready to give the full definition of the concept of uniform fibre bundle. As in the topological case we take a topological base space B, a topological group G, an open covering $\{B_j\}$ of B, and a family $\{g_{jk}\}$ of continuous functions $g_{jk}: B_j \cap B_k \to G$, satisfying the cocycle condition. This time, however, we take as fibre a uniform G-space T, in the sense we have described.

Consider the uniform product $B \times T \times J$, where B has the indiscrete and J the discrete uniformity. We regard the subspace X', defined as before, as a fibrewise uniform space over B using the first projection. I assert that the fibrewise quotient $X = X'/R$ with natural projection π can be given fibrewise uniform structure as follows: a subset D of X^2 is an entourage of X if and only if the inverse image $\pi^{-2}D$ is an entourage of X'. In fact the first two conditions for a fibrewise uniformity are obvious, and the third can be verified as follows.

Let D' be an entourage of X' and let b be a point of B. Since X' is fibrewise uniform there exists a neighbourhood W of b and a symmetric entourage E' of X' such that

$$(X_W'^2 \cap E') \circ (X_W'^2 \cap E') \subset D'.$$

Since the action is uniform there exists a neighbourhood $W' \subset W$ of b and a symmetric entourage F' of X' such that

$$R \circ (X_{W'}'^2 \cap F') \circ R \subset X_{W'}'^2 \cap E'.$$

Then

$$R \circ (X_{W'}'^2 \cap F') \circ R \circ R \circ (X_{W'}'^2 \cap F') \circ R \subset D'.$$

from which the third condition follows at once.

In fact we may regard X as one example of a fibrewise uniform quotient space, another being the maximal fibrewise separated quotient of §15. Note that if $\alpha: X \to Y$ is a fibrewise function, where Y is fibrewise uniform over B, then α is fibrewise uniformly continuous if and only if $\alpha\pi: X' \to Y$ is fibrewise uniformly continuous.

Of course one may drop down to the topological level and regard T as a topological G-space with uniform topology. Then the same structural data can be used to construct a topological fibre bundle over B with fibre T. Clearly this is just the fibrewise topological space associated with the uniform fibre bundle we have just constructed. Examples of uniform fibre bundles include principal bundles and orthogonal sphere-bundles.

The procedure we have just described is functorial in the following sense. Let $\phi: T_1 \to T_2$ be a uniform G-map, where T_1 and T_2 are uniform G-spaces. Then

$$\text{id} \times \phi \times \text{id}: B \times T_1 \times J \to B \times T_2 \times J$$

is fibrewise uniformly continuous, and since X_1 is a fibrewise uniform quotient it follows that the induced fibrewise function $\psi: X_1 \to X_2$ is fibrewise uniformly continuous.

For example let T be a uniform G-space and let $\hat{T}$ be the separated completion of T, with canonical function $\rho: T \to \hat{T}$. Then $\hat{T}$ is a uniform G-space, in the obvious way, and ρ is equivariant. In this case the uniform fibre bundle with fibre $\hat{T}$ can be identified with the fibrewise separated completion of the uniform bundle with fibre T, and the fibrewise function determined by ρ with the canonical function.

Finally, I take this opportunity to make a few remarks about a process which is used a great deal in fibre bundle theory as follows. Let X be a fibrewise set over B and let $\{B_j\}$ be an open covering of B. Suppose that $X_j = p^{-1}B_j$ is equipped with a fibrewise topology over B_j for each index j. Suppose that the intersection $X_j \cap X_k$ is open in both X_j and X_k for each pair of indices j, k. Suppose also that the intersection has the same topology whether it is obtained by restriction from that of X_j or that of X_k.

Then a unique fibrewise topology on X can be defined which restricts to the given topology on X_j for each index j. To be precise, we take the fibrewise basic neighbourhoods in X of a given point $x \in X_b$ to be generated, as a filter, by the neighbourhoods of x in X_j for each index j such that $b \in B_j$. We refer to this process as *fibrewise collation*.

For example, consider the following well-known construction in vector bundle theory. We work in the category $\mathscr{V}_B$ of real, complex or quaternionic finite-dimensional vector bundles over B. If $\mathscr{V}$ denotes the corresponding category of vector spaces we follow Atiyah [2] and describe an endofunctor Φ of $\mathscr{V}$ as *continuous* if the function

$$\Phi_{\#} : \hom(U, V) \to \hom(\Phi U, \Phi V)$$

is continuous, for all $U, V \in \mathscr{V}$. For such functors Φ an endofunctor Φ_B of $\mathscr{V}_B$ can be defined as follows, for all topological spaces B.

At the fibrewise set level we set

$$\Phi_b E = \coprod_{b \in B} \Phi(E_b)$$

for each fibrewise set E, and similarly with fibrewise functions.

Now if $E = B \times T$, for some topological space T, we set

$$\Phi_B E = B \times \Phi T.$$

Moreover if $\alpha: B \times U \to B \times V$ is a homomorphism of vector bundles, then the adjoint $\hat{\alpha}: B \to \hom(U, V)$ of the second projection of α transforms into a function

$$\Phi_\# (\hat{\alpha}): B \to \hom(\Phi U, \Phi V),$$

which is also continuous, and then the adjoint

$$B \times \Phi U \to B \times \Phi V$$

of $\Phi_\# (\hat{\alpha})$ is taken to be $\Phi_B(\alpha)$.

Suppose now that E is a trivial vector bundle over B, so that there exists an isomorphism

$$\alpha: E \to B \times T$$

for some vector space T. Then we can transfer the fibrewise topology of $B \times \Phi T$ to the fibrewise set $\Phi_B E$ through the fibrewise function

$$\Phi_B \alpha: \Phi_B E \to \Phi_B(B \times T) = B \times \Phi T.$$

Moreover the fibrewise topology thus obtained is independent of the choice of trivialization.

Finally consider the general case, where E is locally trivial. Then there exists an open covering $\{B_j\}$ of B such that E is constructed by fibrewise collation from trivial bundles $E_j = E \mid B_j$ for each index j. By what we have already said the transform $\Phi_{B_j}(E_j)$ is defined, as a fibrewise topological space over B_j. Therefore $\Phi_B(E)$ itself is defined, as a fibrewise topological space over B, by fibrewise collation.

Of course the same kind of construction can be employed in the case of multiple functors, whether covariant, contravariant or mixed: details are given in Karoubi [40].

In most applications, if not all, the fibrewise mapping-space can be used as an alternative to fibrewise collation. For example, take the dual of a vector bundle E over B, with field $\mathbf{k}$. This can be defined to be the subspace of $\mathrm{map}_B(E, B \times \mathbf{k})$ for which the fibre over $b \in B$ consists of the linear functionals on the corresponding fibre E_b of E. As in §11 we see that this is equivalent to the result obtained by fibrewise collation, and it has the advantage of applying to fibrewise vector spaces generally.

26. Numerable coverings

Among the many illustrations of fibrewise homotopy in the literature I have chosen one which seems to exemplify particularly well the relation between the fibrewise homotopy theory of a fibrewise topological space, for example the fibrewise Whitehead product structure, and the topology of the base space. We begin with some simple results given in [32] but doubtless well known to anyone who has thought about the subject at all. Deeper results of the same type may be found in [15] and [17].

Proposition (26.1). *Let X and Y be fibrewise spaces over B. Let $\{X_0, X_1\}$ be a numerable covering of X and let $\{Y_0, Y_1\}$ be a family of subsets of Y. Let $\phi_i\colon X_i \to Y_i$ $(i=0, 1)$ be fibrewise maps. Suppose that there exists a fibrewise homotopy*

$$h_t\colon X_0 \cap X_1 \to Y_0 \cap Y_1,$$

where h_0 is given by ϕ_0 and h_1 is given by ϕ_1. Then there exists a fibrewise map $\phi\colon X \to Y$ such that $\phi X_i \subset Y_i$ $(i=0,1)$ and such that there exist fibrewise homotopies

$$f_t\colon X_0 \to Y_0, \qquad g_t\colon X_1 \to Y_1,$$

where $f_0 = \phi_0$, $g_0 = \phi_1$, and where f_1 and g_1 are given by ϕ.

Let $\{\pi, \rho\}$ be a numeration of $\{X_0, X_1\}$, so that $\rho = 0$ away from X_1. Without real loss of generality we may assume h_t to be *bordered*, in the sense of being stationary for $0 \leqslant t \leqslant \frac{1}{4}$ and for $\frac{3}{4} \leqslant t \leqslant 1$. Then ϕ is defined by

$$\phi(x) = \begin{cases} h_{\rho(x)}(x) & (x \in X_0 \cap X_1) \\ \phi_0(x) & (0 \leqslant \rho(x) < \frac{1}{4}) \\ \phi_1(x) & (\frac{3}{4} < \rho(x) \leqslant 1). \end{cases}$$

Also f_t is defined by

$$f_t(x) = \begin{cases} h_{t \cdot \rho(x)}(x) & (x \in X_0 \cap X_1) \\ \phi_0(x) & (0 \leqslant \rho(x) < \frac{1}{4}) \end{cases}$$

and g_t is defined similarly, using π instead of ρ. Since $\rho^{-1}(0, 1) \subset X_0 \cap X_1$, and since X is covered by

$$\{\rho^{-1}(0, 1), \rho^{-1}[0, \tfrac{1}{4}), \rho^{-1}(\tfrac{3}{4}, 1]\}$$

the continuity of ϕ follows at once, and the continuity of f_t and g_t follows similarly.

As an application we prove

Proposition (26.2). *Let X be a fibrewise space over B with projection p. Let $\{X_0, X_1\}$ be a family of subsets of X and let $\{B_0, B_1\}$ be a numerable covering of B such that $pX_i \subset B_i$ $(i=0, 1)$. Suppose that X_i is sectionable* over B_i ($i = 0, 1$) and that $X_0 \cap X_1$ is fibrewise contractible over $B_0 \cap B_1$. Then there exists a section $s: B \to X$ such that $sB_i \subset X_i$ ($i = 0, 1$).

In (26.1) we take X to be B and Y to be X, so that fibrewise maps are sections and fibrewise homotopies are vertical homotopies. Let $s_i: B_i \to X_i$ ($i = 0, 1$) be sections. Since $X_0 \cap X_1$ is fibrewise contractible over $B_0 \cap B_1$ the sections $B_0 \cap B_1 \to X_0 \cap X_1$ determined by s_0, s_1 are vertically homotopic. Hence, by (26.1), there exists a section $s: B \to X$ such that $sB_i \subset X_i$ ($i = 0, 1$), as asserted. Moreover s_0 is vertically homotopic to the restriction of s to B_0 and s_1 is vertically homotopic to the restriction of s to B_1.

Proposition (26.3). *Let X be a fibrewise space over B with projection p. Let $\{X_1, X_0\}$ be a numerable covering of X such that $p \,|\, X_0$, $p \,|\, X_1$ and $p \,|\, X_0 \cap X_1$ are fibrations. Then p itself is a fibration.*

For let $p_i = p \,|\, X_i$ ($i = 0, 1$) and let $p' = p \,|\, X_0 \cap X_1$. Since $\{X_0, X_1\}$ is a numerable covering of X so $\{W(p_0), W(p_1)\}$ is a numerable covering of $W(p)$. Since p_0, p_1 and p' are fibrations we obtain that $P(X_i)$ ($i = 0, 1$) is fibrewise contractible over $W(p_i)$ and that $P(X_0 \cap X_1) = P(X_0) \cap P(X_1)$ is fibrewise contractible over $W(p') = W(p_0) \cap W(p_1)$. Applying (26.2), therefore, we obtain that $P(X)$ is sectionable over $W(p)$ and so p itself is a fibration.

Of course (26.3) can be extended inductively to situations where X admits a finite numerable covering such that the projection of every member of the covering, moreover of every intersection of members of the covering, is a fibre space over B. Following Meiwes [47] we prove

Proposition (26.4). *Let X be a fibrewise space over B with projection p. Let $\{U, V\}$ be a numerable covering of B. Let $\theta, \phi: X \to X$ be fibrewise maps such that $\theta \,|\, X_U$ is fibrewise nulhomotopic over U and $\phi \,|\, X_V$ is fibrewise nulhomotopic over V. Then the composition $\phi\theta$ is fibrewise nulhomotopic over B.*

For let $K: I \times X_U \to X_U$ be a fibrewise homotopy of θ_U into σp_U, and let $L: I \times X_V \to X_V$ be a fibrewise homotopy of ϕ_V into τp_V, where $\sigma: U \to X_U$ and $\tau: V \to X_V$ are sections. Without real loss of generality we may suppose the fibrewise homotopies to be bordered in the same sense as before. Let $\{\pi, \rho\}$ be a numeration of $\{U, V\}$. I assert that $\phi\theta$ is fibrewise homotopic to sp, where $s: B \to X$ is the section given by

$$s(b) = \begin{cases} L(\rho(b), \sigma(b)) & (b \in U \cap V) \\ \phi\sigma(b) & (b \in U \setminus V) \\ \tau(b) & (b \in V \setminus U). \end{cases}$$

To establish this consider the convex hull Δ of the subset $\{(0, 0), (1, 0), (0, 1)\}$ of the real plane $\mathbb{R}^2$. A relative homeomorphism

$$(I \times I, \{1\} \times I) \to (\Delta, (1, 0))$$

is given by $(\xi, \eta) \mapsto (\xi, (1 - \xi)\eta)$. Consider the fibrewise map

$$F: I \times I \times X_{U \cap V} \to X_{U \cap V}$$

which is given for $\xi, \eta \in I$ by

$$F((\xi, \eta), x) = L(\xi, K(\eta, x)) \quad (x \in X_{U \cap V}).$$

Since F is independent of η when $\xi = 1$ there is an induced fibrewise map $G: \Delta \times X_{U \cap V} \to X_{U \cap V}$ which is given on $\partial\Delta \times X_{U \cap V}$ by

$$G((t, 1 - t), x) = L(t, \sigma p(x)),$$

$$G((0, t), x) = \theta K(t, x),$$

$$G((t, 0), x) = L(t, \theta(x)),$$

where $t \in I$ and $x \in X_{U \cap V}$. Thus a fibrewise homotopy $H: I \times X \to X$ of $\phi\theta$ into sp is given for $x \in X_b$ by

$$H(t, x) = \begin{cases} G(t\rho(b), t(1 - \rho(b)), x) & (b \in U \cap V) \\ \phi K(t, x) & (b \in U \setminus V) \\ L(t, \theta(x)) & (b \in V \setminus U). \end{cases}$$

This proves (26.4) and we at once generalize this to

Proposition (26.5). *Let $\{U_i\}$ $(i = 1, \ldots, n)$ be a finite numerable covering of the space B. Let X be a fibrewise space over B. Let $\theta_i: X \to X$ $(i = 1, \ldots, n)$ be fibrewise maps such that the restriction $\theta_i \mid U_i$ is fibrewise nulhomotopic over U_i for each index i. Then the composition $\theta_1 \circ \cdots \circ \theta_n$ is fibrewise nulhomotopic over B.*

We proceed by induction on k, where $1 \leqslant k < n$. We write $V_k = U_1 \cup \cdots \cup U_k$ and make the inductive hypothesis that $(\theta_1 \circ \cdots \circ \theta_k) \mid V_k$ is fibrewise nulhomotopic over V_k; when $k = 1$ this is an assumption in the statement of the proposition. Then $\{U_{k+1}, V_k\}$ is a numerable covering of

$V_{k+1} = V_k \cup U_{k+1}$. Since $\theta_{k+1} \mid U_{k+1}$ is fibrewise nulhomotopic over U_{k+1} we obtain from (26.4) that $(\theta_1 \circ \cdots \circ \theta_{k+1}) \mid V_{k+1}$ is fibrewise nulhomotopic over V_{k+1}. This establishes the inductive step and proves (26.5).

In particular, taking θ_i to be the identity for every index i we obtain

Corollary (26.6). *Let $\{U_i\}$ $(i = 1, \ldots, n)$ be a finite numerable covering of the space B. Let X be a fibrewise space over B. If $X \mid U_i$ is fibrewise contractible over U_i for each index i then X is fibrewise contractible over B.*

Proposition (26.7). *Let $\{U_i\}$ $(i = 1, \ldots, n)$ be a finite numerable covering of the space B. Let $\phi : E \to F$ be a fibrewise map, where E and F are fibrewise spaces over B. If $\phi \mid U_i$ is a fibrewise homotopy equivalence over U_i for each index i then ϕ is a fibrewise homotopy equivalence over B.*

We apply (26.5) with F in place of B and the fibrewise mapping path-space $W_B(\phi)$ in place of X, the projection being ρ_1. We identify the restriction $W_B(\phi) \mid F_i$ with the fibrewise mapping path-space $W_{U_i}(\phi_i)$, where $F_i = F_{U_i}$ and $\phi_i : E_i \to F_i$ is the restriction of ϕ_i. Now $W_{U_i}(\phi_i) = W_B(\phi) \mid F_i$ is fibrewise contractible over F_i, by (23.12), since the projection is a fibrewise fibration and since ϕ_i is a fibrewise homotopy equivalence over U_i. Moreover $\{F_i\}$ $(i = 1, \ldots, n)$ constitutes a finite numerable covering of F, since $\{U_i\}$ is a finite numerable covering of B. Applying (26.6) we see that $W_B(\phi)$ is fibrewise contractible over F and hence that ϕ is a fibrewise homotopy equivalence over B.

This last result can be used to obtain information about Whitehead products. Returning to the situation considered at the end of Chapter IV, let P be a principal G-bundle over B. Let X_i $(i = 1, 2)$ be a non-degenerate G-space, so that $E_i = P_\# X_i$ is a fibrewise non-degenerate space over B. Then for all pointed G-spaces X we have a commutative diagram as shown below, where the upper horizontal is the equivariant Whitehead product and the lower horizontal is the fibrewise Whitehead product, and $E = P_\#(X)$.

$$
\begin{array}{ccc}
\pi_G^*(\Sigma^*(X_1), X) \times \pi_G^*(\Sigma^*(X_2), X) & \longrightarrow & \pi_G^*(\Sigma^*(X_1 \wedge X_2), X) \\
\Big\downarrow {\scriptstyle P_\# \times P_\#} & & \Big\downarrow {\scriptstyle P_\#} \\
\pi_B^B(\Sigma_B^B(E_1), E) \times \pi_B^B(\Sigma_B^B(E_2), E) & \longrightarrow & \pi_B^B(\Sigma_B^B(E_1 \wedge_B E_2), E)
\end{array}
$$

For example take G to be the subgroup $O(q)$ of $O(q + 1)$ $(q \geqslant 2)$ which leaves the first and last basis vectors of $\mathbb{R}^{q+1}$ fixed. We regard S^q as an

$O(q)$-space, through the standard action of $O(q+1)$, and take P to be a principal $O(q)$-bundle over B. We take $X_1 = X_2 = X$ to be S^q, and then $E_1 = E_2 = E$ is the associated bundle with fibre S^q. Let α denote the fibrewise pointed homotopy class of the identity on E. Since the class τ of the switching map $t : E \wedge_B E \to E \wedge_B E$ has degree $(-1)^{q+1}$ on each fibre it follows that in $\pi^B_B(\Sigma^B_B(E \wedge_B E), \Sigma^B_B(E \wedge_B E))$ the sum

(26.8) $\Sigma^B_B \tau + (-1)^q \Sigma^B_B(\alpha \wedge \alpha)$

has degree zero on each fibre. Suppose, therefore, that B admits a numerable covering by n open sets, each of which is contractible in the pointed sense. Then

(26.9) $(\Sigma^B_B \tau + (-1)^q \Sigma^B_B(\alpha \wedge \alpha))^n = 0,$

by (26.5).

Further suppose that the cogroup-like structure of $\Sigma^B_B(E)$ is homotopy-commutative, as is the case if E is itself a reduced fibrewise suspension, in particular if the structural group can be reduced to $O'(q-1)$ as in §7. Then it follows from (26.9) that

$$2^{n-1}\Sigma^B_B \tau + 2^{n-1}(-1)^q \Sigma^B_B(\alpha \wedge \alpha) = 0.$$

Combining this relation with the results at the end of §24 we see that

$$2^{n-1}[\alpha, \alpha] = 2^{n-1}(-1)^q[\alpha, \alpha]$$

in $\pi^B_B(\Sigma^B_B(E \wedge_B E), \Sigma^B_B E)$; hence $2^n[\alpha, \alpha] = 0$ when q is odd. Moreover, $3[\alpha, [\alpha, \alpha]] = 0$ in $\pi^B_B(\Sigma^B_B(E \wedge_B E \wedge_B E), \Sigma^B_B E)$, from the Jacobi identity; hence $[\alpha, [\alpha, \alpha]] = 0$ when q is odd. Further details are given in [31].

27. Fibrewise connectedness

Finally I would like to give a brief discussion of the fibrewise version of connectedness. Consider the case of a fibrewise topological space X over B. One would certainly regard X as fibrewise connected if each of its fibres is connected. However, this condition turns out to be a little too strong for a satisfactory definition. We proceed as follows.

Let us describe X as *reducible* if there exists a partition of X into a pair U_0, U_1 of fibrewise non-empty open sets (see Figure 6). Note that if X is reducible over B, then $X_{B'}$ is reducible over B' for each subspace B' of B. In particular all the fibres of X are disconnected. We note two auxiliary results.

188 *Miscellaneous topics*

Proposition (27.1). *Let $\phi: X \to X'$ be a continuous fibrewise surjection, where X and X' are fibrewise topological over B. If X' is reducible then so is X.*

For if U'_0, U'_1 is a partition of X' into fibrewise non-empty open sets then U_0, U_1 is a partition of X into fibrewise non-empty open sets, where $U_i = \phi^{-1}U'_i$ $(i = 0, 1)$.

Proposition (27.2). *The fibrewise topological space X over B is reducible if and only if there exists a continuous fibrewise surjection $\lambda: X \to B \times \dot{I}$.*

Clearly $B \times \dot{I}$ is reducible and so in one direction this follows at once from (27.1). For the converse suppose that X admits a partition into fibrewise non-empty open sets U_0, U_1. Then a continuous fibrewise surjection $\lambda: X \to B \times \dot{I}$ is given by $\lambda(x) = (p(x), i)$ for $x \in U_i$, $i = 0, 1$.

Definition (27.3). *The fibrewise topological space X over B is fibrewise connected if X_W is irreducible for each open set W of B.*

This is the case, for example, if each of the fibres of X is connected, but it is not necessary for the fibres to be connected. For example, take $X = \{0, 1, 2\}$, with the topology generated by supersets of $\{0\}$, take $B = \{0, 1\}$, with the analogous topology, and take $p(0) = 0, p(1) = p(2) = 1$. Then X is fibrewise connected, although the fibre $\{1, 2\}$ over 1 is not connected; in fact X is fibrewise discrete.

Our two auxiliary results at once imply

Figure 6

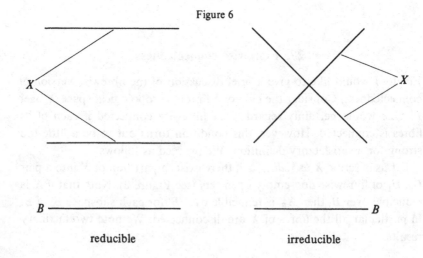

reducible   irreducible

Proposition (27.4). *Let* $\phi: X \to X'$ *be a continuous fibrewise surjection, where X and X' are fibrewise topological over B. If X is fibrewise connected then so is X'.*

Proposition (27.5). *The fibrewise topological space X over B is fibrewise disconnected if and only if X_W is reducible for some open set W of B.*

One of the best-known results of elementary topology characterizes the compact connected subsets of the real line. A fibrewise version of this result† can be obtained as follows. Given real-valued functions $\lambda, \mu: B \to \mathbb{R}$ we write $\lambda \leqslant \mu$ if $\lambda(b) \leqslant \mu(b)$ for all $b \in B$, and in that case we denote by $[\lambda, \mu]$ the fibrewise non-empty subspace of $B \times \mathbb{R}$ consisting of pairs (b, t) such that $\lambda(b) \leqslant t \leqslant \mu(b)$.

Suppose that λ, μ are continuous and that $\lambda \leqslant \mu$. Then a continuous fibrewise surjection $\phi: B \times I \to [\lambda, \mu]$ is given by

$$\phi(b, s) = (b, (1 - s)\lambda(b) + s\mu(b)),$$

and so $[\lambda, \mu]$ is fibrewise compact, fibrewise connected and fibrewise open.

Conversely, suppose that $X \subset B \times \mathbb{R}$ is fibrewise compact, fibrewise connected and fibrewise open. Consider the second component $\alpha: X \to \mathbb{R}$ of the inclusion. Now X is fibrewise closed, as well as fibrewise open, since X is fibrewise compact. For the same reason α is fibrewise bounded and so $\lambda, \mu: B \to \mathbb{R}$ are defined, as continuous functions, where

$$\lambda(b) = \min_{x \in X_b} \alpha(x), \qquad \mu(b) = \max_{x \in X_b} \alpha(x).$$

Clearly $X \subset [\lambda, \mu]$; I assert that equality holds. To see this, first observe that $(b, t) \in X_b$ for $t = \lambda(b)$ and $t = \mu(b)$, since X_b is compact. Suppose, to obtain a contradiction, that $(b, t) \notin X_b$ for some pair (b, t) such that $\lambda(b) < t < \mu(b)$. Since X is closed in $B \times \mathbb{R}$ there exists a neighbourhood W' of b such that $W' \times \{t\}$ does not meet X. Since λ and μ are continuous there exists a neighbourhood $W \subset W'$ of b such that $\lambda(\beta) < t < \mu(\beta)$ for all $\beta \in W$. Now the open sets

$$U_0 = X_W \cap (W \times (-\infty, t)), \qquad U_1 = X_W \cap (W \times (t, \infty))$$

are disjoint and cover X_W, since $W \times \{t\}$ does not meet X. Moreover, if $\beta \in W$ then $(\beta, \lambda(\beta)) \in U_0$, since $\lambda(\beta) < t$, and similarly $(\beta, \mu(\beta)) \in U_1$, so that U_0 and U_1 are fibrewise non-empty. However, X_W is irreducible, since X is fibrewise connected, and so we have our contradiction. We conclude,

---

† This result is suggested by a similar one in Professor Lever's work on fibrewise calculus, which is being prepared for publication.

therefore, that the fibrewise compact, fibrewise connected and fibrewise open subsets of $B \times \mathbb{R}$ are precisely the 'fibrewise intervals' $[\lambda, \mu]$, for continuous functions $\lambda, \mu: B \to \mathbb{R}$ such that $\lambda \leqslant \mu$.

In the literature particular attention is paid to fibrewise topological spaces in which the projection is monotone, meaning that the fibres are compact and connected. However, I do not propose to discuss the particular features of this class of fibrewise topological spaces in the present work.

Exercises

1. Let G be a closed subgroup of the topological group Γ. Suppose that Γ is locally sectionable over Γ/G. Show that Γ is a principal G-bundle over Γ/G.

2. Let X be a fibre bundle over B, where B is contractible. Show that X is trivial as a fibrewise space over B.

3. Let X be a fibre bundle over B. Show that if $\theta, \phi: B' \to B$ are homotopic maps, where B' is any space, then θ^*X and ϕ^*X are equivalent as fibrewise spaces over B'.

4. The base space B is constructed from two copies $\mathbb{R}_1, \mathbb{R}_2$ of the real line by identifying $\alpha \in \mathbb{R}_1$ with $\alpha \in \mathbb{R}_2$ whenever $\alpha \neq 0$. The fibrewise space X over B is constructed from two copies $\mathbb{R}_1 \times \mathbb{R}, \mathbb{R}_2 \times \mathbb{R}$ of the real plane by identifying $(\alpha, \beta) \in \mathbb{R}_1 \times \mathbb{R}$ with $(\alpha, \beta\alpha^{-1}) \in \mathbb{R}_2 \times \mathbb{R}$ whenever $\alpha \neq 0$. Show that X is a line bundle over B, and is non-trivial [suggested by an example in [64]].

5. The base space B consists of the set $(1, 2, 3, 4)$ with the topology generated by the subsets $(1, 2)$, $(1, 2, 3)$, $(1, 2, 4)$. The fibrewise space X over B consists of the fibrewise set $(\pm 1, \pm 2, \pm 3, \pm 4)$, with the projection given by the modulus function and with the topology generated by the subsets $\pm(1)$, $\pm(2)$, $\pm(1, 2, 3)$, $\pm(1, -2, 4)$. Show that X is a fibre bundle over B, and is non-trivial. [In terms of cardinality X is the smallest non-trivial fibre bundle.]

6. The fibrewise uniform space X over B is said to be fibrewise uniformly disconnected if there exists a fibrewise uniformly continuous surjection $\alpha: X_W \to W \times \dot{I}$, where W is a neighbourhood of a point of B and $\dot{I} = \{0, 1\}$ is discrete. Show that X is fibrewise disconnected, in the fibrewise uniform topology, if X is fibrewise uniformly disconnected, and that the converse holds when X is fibrewise compact and fibrewise Hausdorff.

7. Let $\phi: X \to Y$ be a fibrewise uniformly continuous surjection, where X and Y are fibrewise uniform over B. Show that X is fibrewise uniformly disconnected if Y is fibrewise uniformly disconnected.

8. The uniform G-space X is uniformly discontinuous in the sense that if $(g\xi, \xi) \in D$, for $g \in G$ and $x \in X$, then $g\xi = \xi$. Show that X is fibrewise uniformly discrete over X/G.

Note on the literature

The history of fibrewise topology can be traced back to the ideas of Riemann, well over a century ago. However, the modern development of the subject springs from the work of Hurewicz on fibre spaces, in the early thirties, and the work of Whitney on fibre bundles, a little later. Insofar as the results described in the present volume can be found in the previous literature at all it is mainly in publications of the past twenty-five years. Sixty-five of the most relevant are listed in the bibliography: here I would like to make a few comments on some of these.

Whyburn was the first general topologist to adopt what I would regard as the fibrewise viewpoint in several well-known papers. Notably [63] deals with what I have called fibrewise Stone–Čech compactification. The pioneering work of Whyburn was followed up by Cain [11] and others, including Pasynkov [53] and his colleagues in the Soviet Union. Parallel developments occur in categorical topology, for example in the work of Dyckhoff [21], [22] and Johnstone [36], [37].

So far as I am aware it was Booth and Brown [7], [8] who first constructed a satisfactory fibrewise topology for fibrewise mapping-spaces. Their topology, which may be described as a partial map version of the compact-open, is coarser than the one used in the present volume, which is essentially the same as that considered by Niefield in [50]. Recently the Booth–Brown topology has been revived by Lewis [44]. Fibrewise mapping-spaces also appear, as adjoints of the fibrewise topological product, in work on categorical topology.

Fibrewise uniform spaces were also studied by Niefield [49] in a restricted sense where both base space and total space are required to be uniform and the projection is required to be uniformly continuous. An independent attempt [34] of my own was rather less restrictive in that the base space was allowed to be topological rather than uniform. Experience gained with this compromise theory led to the definitions given by §12. There are some parallels with the theory of fields of uniform spaces described by Dauns and Hofmann [13]. Uniform actions, as in §25, arise naturally in topological dynamics (see de Vries [61], for example).

I do not know whether there is a satisfactory fibrewise version of the theory of metric spaces. One possible approach related to the theory of Dauns and Hofmann mentioned above, has been studied by McClendon [45]. Professor Pasynkov tells me that a member of his research group has developed a fibrewise version of proximity theory but I have not seen details of this.

Much has already been written on fibrewise homotopy theory, for example see [30] and [31]. Of course there is also an extensive literature on fibre spaces and fibre bundles. However the particular version of fibrewise homotopy theory given in Chapter IV has not, I believe, been treated before. The recent thesis [65] of Zhang deals with an analogous version of fibre bundle theory. An account of the theory of fibrewise Whitehead products has been given by Eggar [23] and the corresponding theory of fibrewise Samelson products by Berrick [5] while Scheerer [56] and Zhang [65] consider fibrewise Hopf spaces.

I had originally intended to include a section about fibrewise absolute (neighbourhood) retracts but was unable to develop a completely satisfactory theory. An indication of how this might proceed is given in the last chapter of [32]. Fibrewise euclidean neighbourhood retracts have been considered by Dold [18], [19] in his work on fibrewise fixed-point theory.

References

1. M. Arkowitz, The generalized Whitehead product, *Pacific J. Math.* **12** (1962), 7–23.
2. M.F. Atiyah, *K-theory*. Benjamin (New York), 1967.
3. H.J. Baues, *Commutator Calculus and Groups of Homotopy Classes.* Cambridge University Press (Cambridge), 1981.
4. J.C. Becker, On the existence of A_k-structures on stable vector bundles, *Topology* **9** (1970), 367–84.
5. A.J. Berrick, The Samelson ex-product, *Quart. J. Math., Oxford* (2), **27** (1976), 173–80.
6. P.I. Booth, The exponential law of maps I. *Proc. London Math. Soc.* (3), **20** (1970), 179–92.
7. P.I. Booth and R. Brown, On the application of fibred mapping spaces to exponential laws for bundles, ex-spaces and other categories of maps. *Gen. Top. and its Appl.* **8** (1978), 165–79.
8. P.I. Booth and R. Brown, Spaces of partial maps, fibred mapping spaces and the compact-open topology, *Gen. Top. and its Appl.* **8** (1978), 181–95.
9. N. Bourbaki, *Topologie Générale.* Hermann (Paris), 1965.
10. R. Brown, *Some problems of algebraic topology: A study of function spaces, function complexes and FD-complexes.* D.Phil. thesis, Oxford 1961.
11. G.L. Cain, Compactification of mappings, *Proc. Amer. Math. Soc.* **23** (1969), 298–303.
12. D.E. Cohen, Products and carrier theory, *Proc. London Math. Soc.* (3), **7** (1957), 219–48.
13. J. Dauns and K.L. Hofmann, Representations of rings by sections, *Memoirs of Amer. Math. Soc.* **83** (1968).
14. A.S. Davis, Indexed systems of neighbourhoods, *Amer. Math. Monthly* **68** (1961), 886–93.
15. T. tom Dieck, Partitions of unity in homotopy theory, *Comp. Math.* **23** (1971), 153–67.
16. T. tom Dieck, *Transformation Groups.* De Gruyter (Berlin, New York), 1987.
17. A. Dold, Partitions of unity in the theory of fibrations, *Ann. Math.* **78** (1963), 223–55.
18. A. Dold, The fixed point index of fibre-preserving maps, *Inv. Math.* **25** (1974), 281–97.
19. A. Dold, The fixed point transfer of fibre-preserving maps, *Math. Zeit.* **148** (1976), 215–44.
20. J. Dugundji, *Topology.* Allyn and Bacon (Boston), 1965.
21. R. Dyckhoff, Factorisation theorems and projective spaces in topology. *Math. Zeit.* **127** (1972), 256–64.
22. R. Dyckhoff, Projective resolutions of topological spaces, *J. Pure and Appl. Algebra* **7** (1976), 115–19.
23. M.G. Eggar, *Some problems in algebraic topology.* D.Phil. thesis, Oxford, 1970.
24. M.G. Eggar, The piecing comparison theorem, *Indag. Math.* **35** (1973), 320–30.
25. M.G. Eggar, On structure-preserving maps between spaces with cross-sections, *J. London Math. Soc.* (2), **7** (1973), 303–11.
26. R.H. Fox, Shape theory and covering spaces. In Springer *Lecture Notes in Mathematics*, **375** (1974), 71–90.
27. D.R.A. Harvey, *Some topics in algebraic topology.* D.Phil. thesis, Oxford, 1980.
28. M. Henriksen and J.R. Isbell, Some properties of compactifications, *Duke Math. J.* **25** (1958), 83–105.
29. J.R. Isbell, Atomless parts of spaces, *Math. Scand.* **31** (1972), 5–32.
30. I.M. James, Ex-homotopy theory I, *Illinois J. Math.* **15** (1971), 324–37.

31. I.M. James, Alternative homotopy theories, *Enseign. Math.* (2), **23** (1977), 221–47.
32. I.M. James, *General Topology and Homotopy Theory.* Springer (New York), 1984.
33. I.M. James, Fibrewise general topology. In *Aspects of Topology* (Ed. I.M. James and E.R. Kronheimer), Cambridge University Press (Cambridge), 1984.
34. I.M. James, Uniform spaces over a base, *J. London Math. Soc.* (2), **32** (1985), 328–36.
35. I.M. James, Spaces, *Bull. London Math. Soc.* **18** (1986), 529–59.
36. P.T. Johnstone, The Gleason cover of a topos II, *J. Pure and Appl. Algebra,* **22** (1981), 229–47.
37. P.T. Johnstone, Wallman compactification of locales, *Houston J. Math.* **10** (1984), 201–6.
38. P.T. Johnstone, *Topos Theory.* Academic Press (New York), 1977.
39. A. Joyal and M. Tierney, An extension of the Galois theory of Grothendieck, *Memoirs of Amer. Math. Soc.* **309** (1984).
40. M. Karoubi, *K-theory. An Introduction.* Springer (Berlin), 1978.
41. F.C. Kirwan, *Uniform locales.* Part III essay, Cambridge 1981.
42. D. Lever, Continuous families: categorical aspects, *Cahiers de topologie et géométrie différentielle,* **24** (1983), 393–432.
43. D. Lever, Relative topology. In *Categorical Topology, Proc. Conference Toledo, Ohio, 1983.* Heldermann (Berlin), 1984.
44. L.G. Lewis, Jr., Open maps, colimits and a convenient category of fibre spaces. *Topology and its Applications,* **19** (1985), 75–89.
45. J.F. McClendon, Metric families, *Pacific J. Math.* **57** (1975), 491–509.
46. V.A. Matveev and V.M. Ul'yanov, On $\mathcal{F}$-compactifications of mappings, *Russian Math. Surveys,* **37** (1982), no. 2, 227–8.
47. H. Meiwes, On filtrations and nilpotency . . . , *Manuscripta Math.* **39** (1982), 263–70.
48. T.T. Moore, On Fox's theory of overlays, *Fund. Math.* **99** (1978), 205–11.
49. S.B. Niefield, Cartesianness: topological spaces and affine schemes, *J. Pure and Appl. Algebra* **23** (1982), 147–68.
50. R.S. Palais, On the existence of slices for actions of non-compact Lie groups, *Ann. Math.* **73** (1961), 295–323.
51. R.S. Palais, The classification of G-spaces, *Memoirs of Amer. Math. Soc.* **36** (1960).
52. B.A. Pasynkov and L.Yu. Bobkov (in Russian), Extensions to mappings of certain notions and assertions concerning spaces. In *Mappings and Functors.* Moscow State University (Moscow), 1984.
53. A. Pultr, Pointless uniformities, *Comment. Math. Univ. Carolin.* **25** (1984), 91–120.
54. D. Puppe, Homotopiemengen und ihre induzierten Abbildungen, I. *Math. Zeit.* **69** (1967), 299–344.
55 W. Roelcke and S. Dieroff, *Uniform Structures on Topological Groups and their Quotients.* McGraw Hill (New York), 1981.
56. H. Scheerer, On H-spaces over a base space, *Coll. Mat. Barcelona* **36** (1985), 219–28.
57. E.H. Spanier and J.H.C. Whitehead, On fibre spaces in which the fibre is contractible, *Comm. Math. Helv.* **29** (1955), 1–7.
58. N.E. Steenrod, A convenient category of topological spaces, *Michigan Math. J.* **14** (1967), 133–52.
59. A. Strøm, Note on cofibrations, *Math. Scand.* **19** (1966), 11–14.
60. A. Strøm, Note on cofibrations II, *Math. Scand.* **22** (1969), 130–42.
61. J. de Vries, Universal topological transformation groups, *Gen. Top. and its Appl.* **5** (1975), 107–22.
62. A. Weil, *Sur les Espaces à Structure Uniform et sur la Topologie Générale.* Hermann (Paris), 1937.
63. G.T. Whyburn, Compactification of mappings, *Math. Annalen* **166** (1966), 168–74.
64. N.M.J. Woodhouse and L.J. Mason, The Gerosh group and non-Hausdorff twistor spaces, *Nonlinearity* **1** (1988), 73–114.
65. P. Zhang, *Some problems in algebraic topology.* D.Phil. thesis, Oxford, 1987.

Index